The
Origins of
Theoretical
Population
Genetics

The Chicago
History of
Science
and Medicine

Allen G. Debus, Editor

William B. Provine The
Origins of
Theoretical
Population
Genetics

The University of Chicago Press
Chicago and London

International Standard Book Number: 0–226–68465–2
Library of Congress Catalog Card Number: 73–153711

THE UNIVERSITY OF CHICAGO PRESS, CHICAGO 60637
THE UNIVERSITY OF CHICAGO PRESS, LTD., LONDON

To
Doris Marie Provine

Contents

Introduction

We have come together into this hall from various distances, from various states and countries, to discuss the problems of our common interest concerning population genetics. It seems to me that the selection of the agenda for this Twentieth Symposium on Quantitative Biology has been most appropriate and timely, because the importance of population genetics, and its bearing on various other branches of biology, have now become recognized not only by investigators in these branches but also by men in the practical business of breeding and in many other fields of theory and application.[1]

WITH THESE WORDS THE CHAIRMAN OPENED THE COLD SPRING Harbor Symposium of 1955, that year devoted to population genetics. His statement is indicative of the wide recognition that population genetics has gained as an important field of biological research, with implications for other areas of biological interest. Anthropologists, eugenicists, demographers, ecologists, breeders, and others have been much influenced by the impinging ideas of population genetics.

The term "twentieth-century Darwinism" has often been applied to modern population genetics. Although the term may be accurate as a description of the similarity between Darwin's idea of evolution and that of most population geneticists regarding the role of natural selection in the origin of species, it is misleading because it suggests that population genetics developed from Darwin's ideas. The development from Darwin's ideas to population genetics was actually a tortuous one.

The origin of population genetics is perhaps best understood as a product of the conflict between two views of evolution which were eventually synthesized. On one side was

1. *Cold Spring Harbor Symposia on Quantitative Biology* 20 (1955): v.

Darwin's belief in gradual evolution, produced by natural selection acting upon small continuous variations. On the other was Galton's belief in discontinuous evolution, produced by natural selection acting upon large discontinuous variations. Galton thought natural selection was ineffective acting upon the small variations Darwin envisioned. The conflict between these two views began with the publication of the first edition of *The Origin of Species* and did not end until population genetics provided a new, synthetic theory.

In this account, I treat the historical development of the ideas which culminated in the laying of the theoretical foundations of population genetics. The theoretical foundations, sometimes termed "classical" population genetics, were laid between 1918 and 1932 by R. A. Fisher, J. B. S. Haldane, and Sewall Wright. Their work was stimulated in large part by the controversy over the continuity of evolution and the efficacy of natural selection.

I have not included the contributions of the Russian School —Chetverikov, Timofeev-Resovsky, Dubinin, and later Dobzhansky—[2] to the study of natural populations because their work did not merge with "classical" population genetics until after the theoretical foundations were established. Chetverikov did publish in 1927 a very important paper in theoretical population genetics, but by the time this paper was known in England and the United States, the theoretical construct erected by Fisher, Haldane, and Wright had progressed beyond it.

I am greatly indebted to Professors Allen G. Debus and Richard C. Lewontin of the University of Chicago for their careful criticisms and suggestions while I was undertaking this project. I owe special thanks to Professor Lewontin for taking many hours of his time to encourage my work in biology while I was a graduate student in history. Doris Marie

2. For an account of the Russian School, see Mark B. Adams, "The Founding of Population Genetics: Contributions of the Chetverikov School, 1924–1934," *Journal of the History of Biology* 1 (1968): 23–40.

Provine carefully analyzed each chapter, removing deadwood and demanding clarification. I have incorporated so many of her suggestions that she is responsible for a substantial portion of whatever merit this work achieves.

Sewall Wright kindly granted me two lengthy interviews which helped my understanding of his work. I also wish to thank Professor William Coleman of Johns Hopkins University and Mr. Murphy Smith of the American Philosophical Society for their help in guiding me to and through the Bateson papers.

Finally, I am grateful to Mrs. Lina Hood and Mrs. Bernice White for typing services rendered.

1

Darwin's Theory of Natural Selection: The Reaction

When Charles Darwin boarded the Beagle in late 1831 for his famous voyage he took with him volume one of Charles Lyell's *Principles of Geology,* which had been published in 1830. Darwin's teacher John Henslow had recommended the book to him with the admonition not to accept Lyell's views. Like most geologists of the time, Henslow was a catastrophist. He believed the geological history of the earth was progressive, that is, showed significant changes, and was characterized by successive cataclysms with periods of little change in between. Adam Sedgwick, president of the Geological Society in 1831 and Darwin's other teacher of geology, was also a catastrophist. With two catastrophists as his teachers Darwin naturally adopted their general view of geological change. But he did not think catastrophism provided a complete explanation of geological change. He said to a friend, "it strikes me that all our knowledge about the structure of the earth is very much like what an old hen would know of a hundred-acre field, in a corner of which she is scratching." [1]

In the *Principles* Lyell challenged, as James Hutton had before him, the prevalent geological theories of catastrophism and progressionism. Lyell believed the geological history of the earth could be explained by the same agents that were operating at present, given enough time. This was the principle of uniformitarianism. He went further and rejected progressionism. He held that geological forces had made only minor changes in the earth's surface, and not even a cumulation of geological events could cause a major change.

The impact of volume one of Lyell's *Principles* upon Dar-

---

1. Francis Darwin, ed., *The Life and Letters of Charles Darwin,* 3 vols. (London: John Murray, 1887), 2:348.

win's thought was quick and deep. Only twenty days after
the start of the Beagle's voyage, the first ten of which he was
miserably seasick, Darwin landed at St. Iago in the Cape
Verde Islands and was there convinced, as he later said, of
"the infinite superiority of Lyell's views over those advocated
in any other work known to me."[2] This was a remarkable
transformation in such a short period of time.

St. Iago was a perfect place to convince Darwin of Lyell's
belief in gradual geological change. At first glance, the island
might have appeared to be a perfect example of catastrophism.
It had extinct volcanoes and a twenty-foot-thick layer of white
limestone, which had been deposited beneath the water, now
elevated sixty feet above sea level. But Darwin was looking
for evidence of Lyell's theory, and he found it right under the
catastrophists' volcano: the horizontal layers of rock all bent
down into the water in the neighborhood of the volcano, indicating gradual subsidence of the crater.

Lyell's ideas thus gained an auspicious start in Darwin's
mind. He said later that St. Iago "showed me clearly the
wonderful superiority of Lyell's manner of treating geology,
compared with that of any other author, whose works I had
with me or ever afterwards read."[3] Observing geological
patterns during the rest of the Beagle's voyage and reading
volumes two and three of the *Principles* thoroughly convinced
Darwin that Lyell's thesis of gradual change was a true explanation of geological events.

At no time, however, was Darwin convinced of Lyell's ob-
jection to progressive change. He observed too many examples
of the great rearrangements caused by the cumulative effects
of smaller changes. For example, at Port Desire, on the east
coast of South America, he saw beds containing shells from
currently existing species that had been raised to a height of
330 feet above sea level. From his reading of Lyell and his
observations on the voyage of the Beagle, Darwin returned

2. Charles Darwin, *Autobiography,* ed. Nora Barlow (London: Collins, 1958), p. 101.
3. Ibid., p. 77.

to England in 1836 confirmed in the belief that geological change was gradual but progressive.

Ever since his visit to St. Iago at the start of the voyage, Darwin had wanted to write a book on geology. In the ten years after his return he produced three: *The Structure and Distribution of Coral Reefs* (1842), *Geological Observations on the Volcanic Islands* (1844), and *Geological Observations on South America* (1846). All three books embodied the idea of gradual geological change—an idea evidently deep-seated in Darwin's mind.

Darwin considered what he had learned from Lyell in geology to be applicable to biological problems. Speaking of his initial attempt to shed light on the modification of species he later wrote:

> After my return to England it appeared to me that by following the example of Lyell in geology, and by collecting all facts which bore in any way on the variation of animals and plants under domestication and nature, some light might perhaps be thrown on the whole subject.[4]

T. H. Huxley, in his obituary notice for Darwin, commented on the success with which Darwin used Lyell's ideas:

> It is hardly too much to say that Darwin's greatest work is the outcome of the unflinching application to Biology of the leading ideas and the method applied in the "Principles" to geology.[5]

Darwin did not adopt Lyell's ideas about the modification of species, since Lyell's conception of uniformitarianism precluded the possibility of progressive changes in species. What Darwin did apply to the problem of species modification was the same thing he learned from Lyell to apply to geology, namely, the idea of gradual change.

From the beginning of his work on the modification of species Darwin was strongly inclined toward the view of gradual change. When he became convinced that species did

4. Ibid., p. 119.
5. Thomas H. Huxley, *Darwiniana* (New York: D. Appleton, 1896), p. 268.

in fact change, he supposed "that species gradually became modified."[6] The mechanism of species change for which he searched would reflect this attitude.

In July 1837, Darwin began the first of three notebooks on the transmutation of species. By October 1838, he had read Malthus and come to a tentative formulation of the idea of natural selection, the mechanism of species change. Darwin's early vision of natural selection was a simple and clear generalization:

> If variation be admitted to occur occasionally in some wild animals, and how can we doubt it, when we see [all] thousands ⟨of⟩ organisms, for whatever use taken by man, do vary. If we admit such variations tend to be hereditary, and how can we doubt it when we ⟨remember⟩ resemblances of feature and character,—disease and monstrosities inherited and endless races produced (1200 cabbages). If we admit selection is steadily at work, and who will doubt it, when he considers amount of food on an average fixed and reproductive powers act in geometrical ratio. If we admit that external conditions vary, as all geology proclaims, they have done and are now doing,—then, if no law of nature be opposed, there must occasionally be formed races, [slightly] differing from the parent races.[7]

In short, variation exists and is heritable; more organisms are born than can possibly survive on the available supply of food; therefore, those organisms with variations best suited to the environment survive more often and new races are occasionally formed. Thus stated, Huxley's comment seems apt: "How extremely stupid not to have thought of that!"[8] Yet the idea of natural selection involved difficulties which Darwin was never able to overcome fully, which invited the criticism of his friends as well as his opponents, and which were not satisfactorily solved until the rise of population

6. Darwin, *Autobiography,* p. 119.
7. Charles Darwin and Alfred Russel Wallace, *Evolution by Natural Selection* (including "Sketch of 1842," "Essay of 1844," and "Papers of 1858"), ed. Francis Darwin (Cambridge: University Press, 1958), "Sketch of 1842," p. 57. Words in [ ] were erased by Darwin; words in ⟨ ⟩ are editor's insertions.
8. Leonard Huxley, *Life and Letters of Thomas Henry Huxley,* 2 vols. (New York: D. Appleton, 1900), 1:183.

genetics. There were two basic interconnected problems: the character and origin of the variation upon which natural selection acts, and Darwin's assumption of blending inheritance.

Darwin believed that variation was a basic property of species of organisms. In 1842, when he wrote the first sketch of his theory of species modification, he stated at the beginning that

> . . . simple generation, especially under new conditions [when no crossing] causes infinite variation . . . There seems to be no part . . . of body, internal or external, or mind or habits, or instincts which does not vary in some small degree and [often] some to a great amount.[9]

Domestication usually subjected organisms to changed conditions and thus produced more variation than was normally found under natural conditions. But a change of conditions in nature would have the similar effect of causing more variation. New variation was produced each generation.

The variations, as Darwin and most breeders recognized, were of two types. There were sports, large discontinuous variations, relatively rare but sometimes used to good advantage by breeders. The Ancon sheep with short stubby legs was a case of sporting which Darwin mentions on several occasions. Besides sports, there were the less obvious but more pervasive and plentiful minor variations which occurred in every character of the organism. Every species exhibited these minor variations and Darwin believed they were increased when the species was subject to changed conditions. He termed these minor variations "mere variability" or, more often, "individual differences."

Darwin thought both kinds of variation were often inherited. In the very first paragraph of the "Sketch of 1842," speaking of the variations produced by changed conditions, he stated "most of these slight variations tend to become hereditary."[10] Thus variation was a fundamental property of

9. "Sketch of 1842," pp. 41–42.
10. Ibid., p. 41.

species. Under changed conditions variation was bursting out of the population and was mostly heritable.

In the face of so much new variation with each generation, a species could scarcely retain constant characters over a period of generations except by some mechanism which enforced uniformity. To Darwin's mind, blending inheritance supplied this mechanism. Blending of characters may be observed in most sexually reproducing populations and the prevalent opinion in Darwin's time was that the hereditary material itself blended. He adopted that view.[11]

Blending inheritance fit nicely with Darwin's ideas about variation since it kept a species uniform in the face of burgeoning variation. In the "Sketch of 1842," Darwin states that free crossing is a "great agent in producing uniformity in any breed."[12] And in the first edition of the *Origin* he says, "Intercrossing plays a very important part in nature in keeping individuals of the same species, or of the same variety, true and uniform in character."[13] Darwin believed that blending inheritance worked upon both sports and individual differences. He recognized, however, that some sports were prepotent to some degree. They would appear more fully formed in the offspring than predicted by the blending theory. Other sports Darwin recognized as reversions to ancestral characters. Sports of these kinds were dissipated more slowly by blending inheritance but were dissipated nevertheless.

Thus for Darwin sexual reproduction was an agent of uniformity not diversity. He believed more variation occurred in sexually reproducing organisms than in those which reproduced asexually. Otherwise, asexually reproducing organisms, with no blending inheritance to assist, would not be able to exhibit the uniformity all naturalists observed. Darwin thought sexual reproduction was actually more widespread than did his contemporaries because of the vigor it unleashed

11. For a detailed account see Peter Vorzimmer, "Charles Darwin and Blending Inheritance," *Isis* 54 (1963): 371–90.

12. "Sketch of 1842," p. 42.

13. Charles Darwin, *The Origin of Species: A Variorum Text,* ed. Morse Peckham (Philadelphia: University of Pennsylvania Press, 1959), p. 195 (hereafter cited as Peckham).

in the offspring under certain conditions. It followed that variation, fodder for natural selection, was also more widespread but kept in check by interbreeding.

The problem of selection in Darwin's mind was how it operated in the face of blending inheritance. If unchecked, blending would demolish the variation upon which selection acted: "If in any country or district all animals of one species be allowed freely to cross, any small tendency in them to vary will be constantly counteracted." [14] "In man's methodical selection, a breeder selects for some definite object, and free intercrossing will wholly stop his work." [15] For selection to be effective, intercrossing had to be suppressed.

In the case of artificial selection the solution was obvious: the person selecting would isolate from the rest of the population those organisms which exhibited the characters he wanted to perpetuate. Blending inheritance would have no chance to dissipate the new characters. Artificial selection might therefore be quite rapid.

Natural selection, however, presented more difficult problems and Darwin did not reach satisfactory solutions for them until many years later. At first he thought the process of natural selection was similar to that of artificial selection. A beneficial variant might be isolated with a small number of his species. Blending inheritance would then dilute the beneficial character of the variant, but if the isolated group were small enough, then "there would be a chance of the new and more serviceable form being nevertheless in some slight degree preserved." [16] The single beneficial variant Darwin had in mind here was what he had termed a "sport." Yet the discontinuous variant could only lead to a minor change in the isolated segment of the population; and this process fit perfectly with Darwin's belief that evolution occurred gradually. Even when he believed that the primary source of variation was discontinuous, he believed that the evolutionary change of the species was gradual and continuous.

14. "Sketch of 1842," pp. 42–43.
15. Peckham, p. 193.
16. Darwin, "Essay of 1844," p. 198.

The idea of natural selection suggested by Darwin in the "Essay of 1844" soon became untenable for him. Sports were rare, and often one of a kind. For natural selection to proceed, more variation was necessary. Moreover, the geographical isolation postulated as necessary by Darwin occurred rarely. For selection to proceed, the rare variant would have to be isolated with a few members of his species by an unusual occurrence. The combination of rarities did not seem convincing to Darwin, who was looking for the all-pervasive mechanism of species change under nature.

By 1856, when he was writing on a projected exposition of his theory of natural selection, Darwin had found some answers. He was now convinced that sports were too rare to be the primary source of variation; in addition, sports were often infertile and easily swamped by blending inheritance in any but the smallest populations. So he turned instead to the other sort of variation, small but plentiful in every species—individual differences. It is crucial to remember that from the time he wrote the first sentence of the "Sketch of 1842," quoted above, until his death, Darwin believed that many of the individual differences were inherited.

Each generation produced individual differences in each character of a species. The variations tended to occur in every direction, giving natural selection a convenient, if small, handle. For instance, in a species of foxes, some would be born with longer claws than most of the population and some with shorter. If the long claws conferred an advantage, then those foxes would survive better and leave more offspring. Gradually more of the foxes would have longer claws and an evolutionary change would have occurred in the species. Isolation was no longer necessary for evolution in a species; enough variation was produced each generation for selection to operate effectively.

This way of looking at natural selection, that is, as operating upon individual differences, fit perfectly with Darwin's idea of geological change derived from Lyell. With this mechanism evolution was necessarily a gradual and con-

tinuous process; yet great changes could be effected given
enough time. It was basically this theory of species change
which Darwin presented in the first edition of the *Origin of
Species* in 1859.

He did not believe this theory was final. Because of blending inheritance, the theory required that much new heritable
variation be produced each generation; otherwise blending
would destroy the variation upon which natural selection
acted. But the mechanism which produced individual differences was not known to Darwin, as he admitted in the first
Origin:

> I have hitherto sometimes spoken as if the variations—so
> common and multiform in organic beings under domestica-
> tion, and in a lesser degree in those in a state of nature—had
> been due to chance. This, of course, is a wholly incorrect ex-
> pression, but it serves to acknowledge plainly our ignorance
> of the cause of each particular variation.[17]

The critics were quick to seize Darwin's profession of ignorance on the production of variation as a loophole into which
other possibilities could be inserted. Many claimed that the
production of variations was directed and that the variations,
rather than natural selection, determined the direction of
evolution.

In 1868, Darwin finally produced his "Provisional Hypothesis of Pangenesis,"[18] an attempt to supply a theory of heredity
that would account for the production of huge numbers of
heritable individual differences. Basically, the theory stated
that each part of an organism throws off "free and minute
atoms of their contents, that is gemmules."[19] The gemmules
multiply and aggregate in the reproductive apparatus, from

17. Peckham, p. 275.
18. Charles Darwin, *The Variation of Plants and Animals under
Domestication*, 2 vols. (New York: Orange Judd, 1868), vol. 2, chap.
27. For an account of the development of Darwin's thought on pangenesis, see Gerald L. Geison, "Darwin and Heredity: the Evolution
of His Hypothesis of Pangenesis," *Journal of the History of Medicine*
24 (1969): 375–411.
19. Darwin, *Variation*, 2:481.

which they are passed on to the following generations. The theory was designed so that the "direct and indirect" influences of the "conditions of life" might become embodied in the hereditary constitution of the organism. If an organism were affected by the environment, the affected parts would throw off changed gemmules which would be inherited, perhaps causing the offspring to vary in a similar fashion. With his theory of pangenesis to account for the production of individual differences, Darwin's theory of the origin of species was complete.

THE REACTION

Contrary to popular belief, the reaction among most biologists to Darwin's idea of the evolution of species was not strongly adverse, especially after the initial impact of the *Origin*.[20] The idea of evolution was not new, having appeared prominently in the works of Buffon, Erasmus Darwin, Lamarck, and many others. A popular book in England since its publication in 1844 was Robert Chambers's *Vestiges of the Natural History of Creation*,[21] a treatise on organic evolution. The huge amount of evidence for evolution that Darwin had collected was convincingly presented in the *Origin*. Most biologists and many others soon came to believe in it.

Darwin's idea of natural selection, however, did arouse a very strong reaction. The random production of variation, with the relentless elimination of the less fit variants, ran entirely against the prevalent view of "design" in nature. The Reverend Dr. Hodge of Princeton University, echoing the feelings of many, stated that "to ignore design as manifested in God's creation is to dethrone God."[22] That the beauty and harmony of living creatures was the result of chance rather than design was abhorrent to most minds.

20. See Alvar Ellegård, *Darwin and the General Reader: The Reception of Darwin's Theory of Evolution in the British Periodical Press, 1859–1872*, Gothenburg Studies in English, vol. 8 (Goteburg: Elanders Bohtrycheri Aktiebolag, 1958).
21. The *Vestiges* went through eleven editions, about twenty-four thousand copies.
22. Quoted by Andrew Dickson White, *A History of the Warfare of Science and Religion*, 2 vols. (New York: D. Appleton, 1930), 1:79.

Design in nature was not inconsistent with evolution since the unfolding of new organic forms could be seen as resulting from a higher order, from a master plan. But Darwin's idea of natural selection denied this possibility. Thus Darwin was in the curious position of having convinced most scholars that evoluton had occurred but not by the means he envisioned. Most of the serious attacks upon Darwinism centered upon the idea of natural selection.

Many of these attacks were initiated by thinkers who found natural selection abhorrent for nonscientific reasons. The production of variation was a favorite target, and Darwin admitted his weakness in this area. But the attacks were not accompanied by convincing substitutes for the mechanism of evolution. Most of the alternate proposals suggested a nonmaterial force which directed the production of variation and, consequently, the direction of evolution. These theories did not find widespread acceptance.

Out of the assault upon natural selection emerged ideas which were instrumental in the rise of population genetics. Curiously enough, this criticism came from two of Darwin's staunchest supporters and admirers, Thomas H. Huxley and Francis Galton. Darwin firmly believed that individual differences were the significant source of variation, that natural selection was the motive agent, and that the movement of evolution was both gradual and continuous. Huxley and Galton challenged Darwin's emphasis upon individual differences. They suggested instead that selection primarily utilized discontinuous variations, or sports, and in consequence that evolution might proceed rapidly and by discontinuous leaps. The criticism and suggestions offered by Huxley and Galton are important, for the division between those who believed in the natural selection of individual differences and continuous evolution and those who believed in a mutation theory and discontinuous evolution was the common thread in the development of population genetics.

THOMAS H. HUXLEY AND "NATURA NON FACIT SALTUM"

Thomas H. Huxley was perhaps the most articulate and vigorous of the early champions of Darwin's ideas. He once said, "I am Darwin's bull-dog," and many of Darwin's critics suffered under his onslaughts. Yet he was among the very first to criticize Darwin's treatment of variation and continuity in evolution. Immediately upon reading the first edition of the *Origin*, Huxley wrote Darwin that "you have loaded yourself with an unnecessary difficulty in adopting *Natura non facit saltum* so unreservedly." [23] The unnecessary difficulty was that the gaps between existing species and in the geological record could not easily be explained if natural selection operated only upon individual differences, because intermediate forms would be expected. Such gaps were to be expected, however, if the raw material for natural selection were discontinuous variations, or what Huxley termed "saltations."

Five months prior to the publication of the *Origin*, Lyell told Huxley that the transmutation theory could not account for the distinct gaps between species, living and fossil. Huxley's reply in a letter of 25 June 1859 states clearly the position he held until the end of his life.

The fixity and definite limitation of species, genera, and larger groups appear to me to be perfectly consistent with the theory of transmutation. In other words, I think *transmutation* may take place without transition.

Suppose that external conditions acting on species *A* give rise to a new species, *B;* the difference between the two species is a certain definable amount which may be called *A-B*. Now I know of no evidence to show that the interval between the two species must necessarily be bridged over by a series of forms; each of which shall occupy, as it occurs, a fraction of the distance between *A* and *B*. On the contrary, in the history of the Ancon sheep, and of the six-fingered Maltese family, given by Réaumur, it appears that the new form appeared at once in full perfection.

I may illustrate what I mean by a chemical example. In an organic compound, having a precise and definite composition, you may effect all sorts of transmutations by substituting an atom of one element for an atom of another element. You may in this way produce a vast series of modifications—but each modification is definite in its composition, and there are no transitional or intermediate steps between one definite com-

23. T. H. Huxley to C. Darwin, 23 November 1850, in L. Huxley, *Life and Letters of Thomas Henry Huxley,* 1:189.

pound and another. I have a sort of notion that similar laws of definite combination rule over the modifications of organic bodies, and that in passing from species to species "Natura fecit saltum." [24]

That natural selection acted upon saltations, instead of individual differences, was for Huxley precisely what harmonized the theory of natural selection and the evidence of geology. Indeed, he believed that the transmutation of species was the natural extension of Lyell's uniformitarianism.

Huxley believed, as did Darwin, that selection played a decisive role in evolution. Speaking of his own two examples, Ancon sheep and hexadactyl humans, Huxley states that in one "a race was produced, because, for several generations, care was taken to select both parents of the breeding stock from animals exhibiting a tendency to vary in the same direction; while, in the other, no race was evolved, because no such selection was exercised." [25] Darwin would never have admitted that natural selection could act so quickly, in "several" generations.

Huxley believed saltations to be more stable and less susceptible to the effects of blending inheritance than did Darwin: "If a variation which approaches the nature of a monstrosity can survive thus forcibly to reproduce itself, is it not wonderful that less aberrant modifications should tend to be preserved even more strongly." [26] These "less aberrant modifications" Darwin would never have considered individual differences, since Ancon sheep are given as an example. Instead the variations referred to by Huxley are more like the "discontinuous variations" later described by William Bateson and called "sports" by Darwin. Huxley's idea of evolution appears to have more in common with de Vries's and Bateson's than with Darwin's.

24. T. H. Huxley to Sir Charles Lyell, 25 June 1859, ibid., pp. 185–86. Considering recent knowledge of the mutations in DNA, the hereditary material, Huxley's last paragraph seems to be a remarkable anticipation of modern genetical thought.
25. T. H. Huxley, "Darwin on the Origin of Species," *Westminster Review*, n.s., 17 (1860): 550.
26. Ibid., p. 549.

FRANCIS GALTON, REGRESSION, AND DISCONTINUOUS EVOLUTION

Francis Galton greatly admired his cousin Charles Darwin, and certainly Darwin influenced Galton's scientific researches. Four years after Darwin's death Galton had this to say:

> Few can have been more profoundly influenced than I was by his publications. They enlarged the horizon of my ideas. I drew from them the breath of a fuller scientific life, and I owe more of my later scientific impulses to the influences of Charles Darwin than I can easily express. I rarely approached his genial presence without an almost overwhelming sense of devotion and reverence, and I valued his encouragement and approbation more perhaps, than that of the whole world besides.[27]

But Galton's brilliant and acute mind scarcely allowed him to agree with Darwin when his own researches conflicted with his cousin's ideas. Galton disagreed most with Darwin on the issues of hereditary variation and continuity in evolution. Huxley probably influenced Galton on these issues, but no direct evidence for this is available.

In 1869, Galton published his book *Hereditary Genius,* a preliminary attempt to analyze the inheritance of genius. He was certain that genius was a hereditary trait and that a genius had a distinctly superior intellect compared to other humans. The spectrum of human intelligence was not uniform, and each type of intelligence seemed to be stable. In the concluding chapter, in a passage which was often quoted by his contemporaries, Galton explained what he meant by stability of types and how that was connected with evolution:

> I will now explain what I presume ought to be understood, when we speak of the stability of types, and what is the nature of the changes through which one type yields to another . . . . It is shown by Mr. Darwin, in his great theory of *The Origin of Species,* that all forms of organic life are in some sense convertible into one another, for all have, according to his views, sprung from common ancestry, and therefore A and B having both descended from C, the lines of descent might be remounted from A to C, and redescended from C to B.

27. From speech of Francis Galton delivered at the Royal Society, 1886. Recorded in Karl Pearson, *The Life, Letters, and Labours of Francis Galton,* 3 vols. (Cambridge: Cambridge University Press, 1914–30), 2:201 (hereafter cited as *Galton*).

Yet the changes are not insensible gradations; there are many, but not an infinite number of intermediate links; how is the law of continuity to be satisfied by a series of changes in jerks? The mechanical conception would be that of a rough stone, having, in consequence of its roughness, a vast number of natural facets, on any one of which it might test in "stable" equilibrium. That is to say, when pushed it would somewhat yield, when pushed much harder it would again yield, but in a less degree; in either case, on pressure being withdrawn it would fall back into its first position. But, if by a powerful effort the stone is compelled to overpass the limits of the facet on which it has hitherto found rest, it will tumble over into a new position of stability, whence just the same proceedings must be gone through as before, before it can be dislodged and rolled another step onwards. The various positions of stable equilibrium may be looked upon as so many typical attitudes of the stone, the type being more durable as the limits of its stability are wide. We also see clearly that there is no violation of the law of continuity in the movements of the stone, though it can repose in certain widely separated positions.[28]

Galton at this time already believed that evolution proceeded by discontinuous leaps, a belief he later expressed in more detail.

When writing *Hereditary Genius* Galton based his comments about heredity upon Darwin's hypothesis of pangenesis. Galton's theory of stability of types did not, however, seem to fit with pangenesis, which Darwin had developed primarily to account for the production of hereditary individual differences. Galton therefore decided to test pangenesis experimentally. Darwin had stated that the gemmules "circulate freely throughout the system" and that "the gemmules in each organism must be thoroughly diffused; nor does this seem improbable considering their minuteness, and the steady circulation of fluids throughout the body." [29] Assuming Darwin meant that the gemmules circulated in the blood stream of higher animals, Galton embarked, with Darwin's encouragement, upon an attempt to transfuse the blood of different varieties of rabbits. If the offspring of the

28. Francis Galton, *Hereditary Genius* (reprint of 1892 edition, New York: Meridian Books, 1962), pp. 421–22.
29. Darwin, *Variation,* 2:448, 454.

transfused rabbits were mongrelized, Darwin's theory would
be proved.

The first experiments were negative. The offspring of the
transfused rabbits were in no way affected. After improving
the technique of transfusion, Galton attempted another series of experiments on rabbits, all negative. On 30 March
1871, Galton read a paper before the Royal Society in which
he stated that "the doctrine of Pangenesis, pure and simple,
as I have interpreted it, is incorrect." [30]

Darwin reacted by writing a letter to *Nature* published on
27 April. Galton's interesting experiments, he said, proved
little. Nowhere had he stated that the gemmules must travel
in the blood stream, so Galton's proof that the gemmules
could not be in the blood did not destroy the theory of pangenesis. Darwin admitted that when Galton was performing
the experiments he "did not sufficiently reflect on the subject, and saw not the difficulty of believing in the presence
of gemmules in the blood." [31]

Galton replied with a letter to *Nature* published on 4 May.
After showing that Darwin's ambiguous language in his
statement of pangenesis might easily be interpreted to mean
that the gemmules would be in the blood stream, Galton
concluded:

I do not much complain of having been sent on a false
quest by ambiguous language, for I know how difficult it is
to put thoughts into accurate speech, and, again, how words
have conveyed false impressions on the simplest matters from
the earliest times. Nay, even in the idyllic scene which Mr.
Darwin has sketched of the first invention of language,
awkward blunders must of necessity have occurred. I refer
to the passage in which he supposes some unusually wise
ape-like animal to have first thought of imitating the growl
of a beast of prey so as to indicate to his fellow-monkeys the
nature of expected danger. For my part, I feel as if I had just

30. Francis Galton, "Experiments in Pangenesis by Breeding from
Rabbits of a Pure Variety into Whose Circulation Blood Taken from
Other Varieties Had Been Infused," *Proceedings of the Royal Society*
19 (1871): 404.

31. Darwin to *Nature*, 27 April 1871, reprinted in Pearson, *Galton*,
2:163.

been assisting at such a scene. As if, having heard my trusted leader utter a cry, not particularly well articulated, but to my ears more like that of a hyena than any other animal, and seeing none of my companions stir a step, I had, like a loyal member of the flock, dashed down a path of which I had happily caught sight, into the plain below, followed by the approving nods and kindly grunts of my wise and most respected chief. And now I feel, after returning from my hard expedition, full of information that the suspected danger was a mistake, for there was no sign of a hyena anywhere in the neighborhood. I am given to understand for the first time that my leader's cry had no reference to a hyena in the plain, but to a leopard somewhere up in the trees; his throat had been a little out of order—that was all. Well, my labour has not been in vain; it is something to have established the fact that there are no hyenas in the plain, and I think I see my way to a good position to look out for leopards among the branches of the trees. In the meantime, Vive Pangenesis![32]

Galton's dissatisfaction with Darwin's theory of heredity is visible beneath his humor. Pangenesis was not the "leopard among the branches" that Galton now sought.

Although he considered Darwin his "wise and most respected chief," Galton's ideas on heredity and evolution were to be sufficiently different from those of Darwin that thirty-five years after the pangenesis experiments some of the most outspoken anti-Darwinists claimed to trace their intellectual heritage to Galton. Curiously, some of the most respected Darwinists of that time also claimed Galton as their inspiration. The reasons for this incongruous situation are discussed in chapter 2.

Although Galton knew he had not definitely disproved Darwin's theory of pangenesis, he began to search for a theory of heredity which would harmonize with his belief in the discontinuity in variation. By 1875 his theory was taking definite shape. The hereditary qualities, instead of being imbedded in gemmules which were formed in all parts of an organism, were concentrated in the reproductive organs. The germ plasm, or "stirp," was continuously inherited from generation to generation with little alteration. One corollary was

32. Galton to *Nature*, 4 May 1871, ibid., p. 165.

that hereditary variations were caused by alterations of the "stirp" and were quite distinct. Another corollary was that acquired characters were unlikely to affect the germ plasm, as would happen under the hypothesis of pangenesis.

With his own theory of heredity in hand, Galton again examined the problem of continuity in evolution. There can be no doubt that he advocated the idea of evolution by discontinuous leaps. In 1894, after discussing some instances of sports described by Darwin in *Variation of Plants and Animals under Domestication*, Galton states that "many, if not most breeds, have had their origin in sports." He goes on to say:

Notwithstanding a multitude of striking cases of the above description collected by Darwin, the most marked impression left on his mind by the sum of all his investigations was the paramount effect of the accumulation of a succession of petty differences through the influence of natural selection. This is certainly the prevalent idea among his successors at the present day, with the corollary that the Evolution of races and species has always been an enormously protracted process. I have myself written many times during the last few years in an opposite sense to this, more especially in three works: *Natural Inheritance*, 1889, in *Finger Prints*, 1892, and in the preface to a reprint of *Hereditary Genius*, 1892.[33]

In *Natural Inheritance*, under a section entitled "Evolution not by Minute Steps Only," Galton stated:

The theory of Natural Selection might dispense with a restriction, for which it is difficult to see either the need or the justification, namely, that the course of evolution always proceeds by steps that are severally minute, and that become effective only through accumulation.[34]

And in *Finger Prints*:

The progress of evolution is not a smooth and uniform progression, but one that proceeds by jerks, through successive

33. Francis Galton, "Discontinuity in Evolution," *Mind*, n.s., 3 (1894): 365, 366.
34. Francis Galton, *Natural Inheritance* (London: Macmillan, 1889), p. 32.

'sports' (as they are called), some of them implying considerable organic changes; and each in its turn being favored by
Natural Selection.[35]

Why did Galton break so decisively with Darwin on the
issue of discontinuity in evolution? Two reasons are Galton's
beliefs in the principle of regression and in the stability of
sports. The phenomenon of regression is clear: In a population whose general characters remain constant over a period
of generations, each character nevertheless exhibits some
variability each generation. Yet the range of this variability
does not change from generation to generation. Thus the
exceptional members of the population cannot produce even
more exceptional offspring, on an average, or else the range
of variability of the character in question would expand
markedly. Indeed, since those with average characters produce some with exceptional, then the exceptional members
must tend to produce less exceptional offspring. In short,

. . . the ordinary genealogical course of a race consists in a
constant outgrowth from its centre, a constant dying away at
its margins, and a tendency of the scanty remnants of all
exceptional stock to revert to that mediocrity, whence the
majority of their ancestors originally sprang.[36]

Though not a serious mathematician, Galton was eager to
quantify the general laws he observed. Regression fascinated
him and he attempted to gather data from which to derive a
quantitative law. His data covered the inheritance of size in
the sweet pea, and of stature, eye color, temper, artistic faculty, and disease in man. Between 1877 and 1888, Galton
published several papers on regression and in 1889 published
his book *Natural Inheritance*.

The first six chapters of this book contained discussions of
heredity, organic stability, and statistical methods. In the
succeeding chapters Galton launched into an analysis of his
data. First he treated stature in man:

35. Francis Galton, *Finger Prints* (London: Macmillan, 1892), p. 20.
36. Francis Galton, "Typical Laws of Heredity," *Journal of the
Royal Institution* 8 (1875–77): 298.

However paradoxical it may appear at first sight, it is theoretically a necessary fact, and one that is clearly confirmed by observation, that the Stature of the adult offspring must on the whole, be more *mediocre* than the stature of their parents; that is to say, more near to the M [median or mid-stature] of the general Population.[37]

This of course was the phenomenon of regression.

Galton adopted the following scheme for a quantitative measure of regression. Suppose groups I and II are chosen from a population with a median measure M of some character. Then the median measure of the character in group I may be expressed as $M \pm D$, and in group II as $M \pm kD$. The quantity k Galton defined as the regression of group II on group I with respect to the chosen character. He often expressed k in other words, saying it was the regression from the group I character to the group II character. Note that when Galton calculates the regression between different generations he must assume that the median M of the population stays constant from generation to generation.

With his data on stature Galton first converted all female heights to male heights so he could find the average (mid-) height of a group of mixed sexes. He then found that the average regression of mid-filial stature upon mid-parental stature was about $\frac{3}{5}$, but he later substituted the value $\frac{2}{3}$

because the data seemed to admit of that interpretation also, in which case the fraction of two-thirds was preferable as being the more simple expression. . . .

This value of two-thirds will therefore be accepted as the amount of Regression, on the average of many cases, from the Mid-Parental to the Mid-Filial stature, whatever the Mid-Parental stature may be.[38]

In the next paragraph, it becomes clear why Galton considered $\frac{2}{3}$ the "more simple expression." He wanted to calculate the mid-filial regression on a single parent, but his data were insufficient to calculate the value directly. He adopted the following argument:

37. Galton, *Natural Inheritance,* p. 95.
38. Ibid., p. 98.

As the two parents contribute equally, the contribution of either of them can only be one half of that of the two jointly; in other words, only one half of that of the Mid-Parent. Therefore the average Regression from the Parental to the Mid-Filial Stature must be the one half of two-thirds, or one-third.

The fraction $\frac{3}{5}$ would not have fit so nicely in this calculation.

Galton summed up his discussion of the law of regression in stature:

> The law of Regression in respect to Stature may be phrased as follows; namely, that the Deviation of the Sons from P [the median stature of the general population] are, on the average, equal to one-third of the deviation of the Parent from P, and in the same direction. Or more briefly still:—If $P + (\pm D)$ be the Stature of the Parent, the Stature of the offspring will on the average be $P + (\pm \frac{1}{3} D)$.[39]

In the succeeding chapters and appendixes Galton showed that according to his data the same law of regression held for human eye color, the artistic faculty, consumption, and size in sweet peas. He was confident that the law of regression was a theoretical necessity and would be found to hold for nearly all organisms.

Using the basic law of regression of son on father, Galton calculated the regression between more distant relatives than father and son. Since the regression of the son on the father was $\frac{1}{3}$, and that of the father on his father was also $\frac{1}{3}$, Galton deduced that the regression of the son on the grandfather was $\frac{1}{3} \times \frac{1}{3} = \frac{1}{9}$. He similarly derived the regression to be expected between other relatives. For example, to find the regression of nephews on uncles he reasoned: "a Nephew is the son of a Brother, therefore in this case we have [the regression] $\frac{1}{3} \times \frac{2}{3} = \frac{2}{9}$."[40]

Another derivation Galton made from the law of regres-

---

39. Ibid., p. 104.
40. Ibid., p. 132. Galton's method here is fallacious. It would hold, among other conditions, only if the regressions were entirely independent, which they are not. Pearson has pointed this out in *Galton,* 3A:24.

sion was his later named law of ancestral heredity.[41] He wanted a measure of the separate contribution of each ancestor to a particular character in the offspring. With considerable hand waving [42] he derived the following result:

The influence, pure and simple, of the Mid-Parent may be taken as $\frac{1}{2}$, and that of the Mid-Grand-Parent as $\frac{1}{4}$, and so on. Consequently, the influence of the individual Parent would be $\frac{1}{4}$, and of the individual Grand-Parent $\frac{1}{16}$, and so on. It would, however, be hazardous on the present slender basis, to extend this sequence with confidence to more distant generations.[43]

A few pages later Galton stated the law in a different form: "each unit of peculiarity in each ancestor taken singly, is reduced in transmission according to the following average scale;—a Parent transmits only $\frac{1}{4}$, and a Grand-Parent only $\frac{1}{16}$." [44]

Although Galton derived this law from data concerning stature in man, a continuously varying character, he believed it was applicable to nonblending inheritance, such as eye color in humans. Since a parent could not contribute an eye which was one-quarter blue, Galton treated the total heritage as being represented by percentages of the offspring. Thus the eye color of each parent determined on the average the eye color of one-quarter of the offspring. Similarly, the eye color of each grandparent determined the eye color of one-sixteenth of the offspring. Galton used this formulation in his treatment of eye color in chapter 8 of *Natural Inheritance*.

If Galton's derivations of his regression coefficients and his law of ancestral heredity were questionable, he nevertheless opened the door to a statistical analysis of correlations of characters, an analysis which was to have immense influence

41. Pearson named Galton's law of ancestral contributions "Mr. Galton's Law of Ancestral Heredity" in 1898. This development will be treated in the next chapter.
42. See R. G. Swinburne, "Galton's Law—Formulation and Development," *Annals of Science* 21 (1965): 15–31.
43. Galton, *Natural Inheritance,* p. 136.
44. Ibid., p. 138.

upon evolutionary thought. The biometricians were later to point to *Natural Inheritance* as the starting point of biometry.

The implications of regression and the law of ancestral heredity for evolution seemed obvious to Galton. Selection was ineffective in the face of regression "because an equilibrium between deviation and regression will soon be reached, whereby the best of the offspring will cease to be better than their own sires and dams."[45] The extremes which selection caused would quickly be brought back to the center by the action of regression.

Galton made a clear-cut distinction between "sports" and "variations proper," or "mere variations":

> The same word 'variation' has been indiscriminately applied to two very different conceptions, which ought to be clearly distinguished: the one is that of the 'sports' just alluded to, which are changes in the position of organic stability, and may, through the aid of Natural Selection, become fresh steps in the onward course of evolution; the other is that of Variation proper, which are merely strained conditions of a stable form of organisation, and not in any way an overthrow of them. Sports do not blend freely together; variations proper do so. Natural Selection acts upon variations proper, just as it does upon sports, by preserving the best to become parents, and eliminating the worst, but its action upon mere variation can, as I conceive, be of no permanent value to evolution, because there is a constant tendency in the offspring to 'regress' towards the parental type.[46]

Looking at the effects of blending inheritance and regression, Galton decided that sports must be the only effective source of evolutionary variation. Darwin, looking at blending inheritance, decided just the opposite—that sports could play no role in evolution. Individual differences must be the effective source of variation for natural selection. Darwin was aware of prepotency but did not believe it was widespread enough to keep sports from being obliterated by blending inheritance. Galton believed that sports were actually quite stable.

45. Galton, *Hereditary Genius*, p. 34.
46. Galton, *Finger Prints*, p. 20.

Here then is the basic setting out of which population genetics grew. On the one hand is Darwin's view of gradual
and continuous evolution; on the other is Galton's view of
abrupt and discontinuous evolution. There were others besides Galton and Huxley who believed in evolution by jumps.
Mivart, von Kölliker, and Nägeli were among these, but they
all believed in a nonmaterial directive agency guiding the
production of large mutations. In 1894 Galton, speaking of
his own ideas of discontinuous evolution, was able to say:
"These briefly are the views that I have put forward in various publications during recent years, but all along I seemed
to have spoken to empty air. I never heard nor have I read
any criticism of them, and I believed they had passed un-
heeded and that my opinion was in a minority of one." [47]

Yet within a year a widely publicized controversy arose
about whether evolution was discontinuous or not, and the
combating schools both traced their heritage to Galton. This
controversy was the prelude to the well-known battle between the Mendelians and biometricians. The origins of that
struggle began well before the rediscovery of Mendelian inheritance in 1900.

47. Galton, "Discontinuity in Evolution," p. 369.

2 Background to the Conflict between Mendelians and Biometricians

THE WIDELY PUBLICIZED CONFLICT BETWEEN THE MENDELIANS and biometricians, which arose soon after the rediscovery of Mendel's work in 1900, influenced the development of population genetics. The conflict caused a split between those who advocated Mendel's theory of heredity and those who advocated Darwin's theory of natural selection. If the Mendelians had worked with, instead of against, the biometricians, the synthesis of Mendelian inheritance and Darwinian selection into a mathematical model, later accomplished by population genetics, might have occurred some fifteen years earlier.

To say the conflict was between the Mendelians and biometricians is misleading, since the basic disagreement was recognized by both parties well before the rediscovery of Mendelian inheritance. The real problem was whether evolution proceeded in general by natural selection operating upon small variations, as Darwin believed, or by discontinuous leaps, as both Huxley and Galton believed. The biometricians supported Darwinian evolution and the Mendelians supported discontinuous evolution.

To understand the background of the conflict, up to the rediscovery of Mendelism, a knowledge of the powerful personalities involved and their interactions is necessary. Although he remained aloof during the conflict proper, Galton was deeply involved, for both sides claimed him as one of their own. The biometricians looked upon Galton as the founder of their new science, and the Mendelians saw him as the father of the theory of evolution by discontinuous leaps. Biometricians Karl Pearson and W. F. R. Weldon and arch-Mendelian William Bateson fought for their ideas with vigor. The intensity of their disagreement generated such strong personal antagonisms that collaboration, which might have been very fruitful, was virtually impossible.

Karl Pearson: A Sketch of His Early Life

Karl Pearson was born in London in 1857. His father, William Pearson, was a barrister with a strong interest in history. The younger Pearson later said that his father labored diligently on his legal work and "only in the vacations did we really see him; then he was shooting, fishing, sailing with a like energy which astonished me even as an active boy."[1] Pearson resembled his father in the great energy and diligence he focused on his work.

In 1875 Pearson entered King's College, Cambridge, on a scholarship. He graduated with mathematical honors in 1879 and immediately left for Germany, where he studied in Heidelberg and Berlin. In 1880 he returned to London and was called to the Bar in 1881. In 1884, at the age of twenty-seven, he assumed the chair of Applied Mathematics and Mechanics at University College, London, formerly occupied by William Kingdon Clifford.

Pearson himself has described his unusual mixture of intellectual activities during these years:

In Cambridge I studied Mathematics under Routh, Stokes, Cayley, and Clerk Maxwell—but wrote papers on Spinoza. In Heidelberg I studied Physics under Quinke, but also Metaphysics under Kuno Fischer. In Berlin I studied Roman Law under Bruns and Mommsen, but attended the lectures of Du Bois Reymond on Darwinism. Back at Cambridge I worked in the engineering shops but drew up the schedule in Mittel- and Althochdeutsch for the Medieval Languages Tripos. Coming to London, I read in Chambers in Lincoln's Inn, drawing up bills of sale, and was called to the Bar, but varied legal studies by lecturing on Heat at Barnes, on Martin Luther at Hampstead and on Lassalle and Marx on Sundays at revolutionary clubs around Soho. Indeed, I contributed to the Socialist Song Book hymns which I believe are still chanted.[2]

1. Address of Karl Pearson, in *Speeches Delivered at a Dinner Held in University College, London, in Honour of Professor Karl Pearson, 23 April 1934* (privately printed, Cambridge: Cambridge University Press, 1934), p. 20.
2. Ibid.

After returning from Germany, Pearson gave many lectures on German culture, dealing especially with the life and times of Martin Luther. He wrote a treatise in German on engravings of Jesus Christ during the Middle Ages, composed a nineteenth-century passion play, produced reviews on the works of Spinoza, and wrote a large number of other letters, articles, and reviews. In addition, he published several very technical papers, for example, "On the Motion of Spherical and Ellipsoidal Bodies in Fluid Media." [3] He also assumed the difficult task of editing Clifford's *Common Sense of the Exact Sciences* and Isaac Todhunter's *A History of the Theory of Elasticity and of the Strength of Materials from Galilei to the Present Time*. Both of these important works required considerable effort to complete. Besides his writing, Pearson devoted much time to his professional duties, lecturing on geometry and mechanics. Students found him stimulating.

Pearson was an intelligent young man. He knew it, and was quick to criticize the incompetence of others. An example was his attack on an exhibition in 1883 at the British Museum celebrating the three hundredth anniversary of Martin Luther's birth. When others reacted to his criticism, he engaged in a number of literary duels, brandishing sharp-edged rhetoric. Henry Bradshaw, Pearson's most respected and admired teacher, wrote him the following letter concerning these exchanges:

I have not the slightest wish to defend the Museum ignorance. But . . . when a man who might by his own deeper knowledge help to make such an exhibition very much more interesting and instructive wastes his energies in writing to the *Athenaeum* as you do, it naturally produces the impression that his main object is to let the world see how much more he knows of the subject than the idiots to whose care he says these treasures are entrusted. Those who know you know also that that is not the object you have in view, but it is a pardonable inference for ordinary people to draw. Everything you write about this shows such an extraordinary absence of

3. *Quarterly Journal of Pure and Applied Mathematics* 20 (1883): 60–80.

wisdom (by which I don't mean knowledge or cleverness, both of which are abundantly shown). . . .[4]

Pearson allowed this letter to be published in a memoir of Bradshaw in 1888, which indicates he took the criticism to heart. He was still quick, however, to discredit prime examples of sloppy thinking. Later he assigned significant portions of William Bateson's thought to this category.

In 1889, Pearson was much influenced by reading Galton's *Natural Inheritance*. Looking back in 1934, Pearson quoted from Galton's Introduction:

"This part of the enquiry may be said to run along a road on a high level, that affords wide views in unexpected directions, and from which easy descents may be made to totally different goals to those we have now to reach."

Pearson went on to say:

"Road on a high level," "wide views in unexpected directions," "easy descents to totally different goals"—here was a field for an adventurous roamer! I felt like a buccaneer of Drake's days—one of the order of men "not quite pirates, but with decidedly piratical tendencies," as the dictionary has it! I interpreted that sentence of Galton to mean that there was a category broader than causation, namely correlation, of which causation was only the limit, and that this new conception of correlation brought psychology, anthropology, medicine and sociology in large parts into the field of mathematical treatment. It was Galton who first freed me from the prejudice that sound mathematics could only be applied to natural phenomena under the category of causation. Here for the first time was a possibility—I will not say a certainty of reaching knowledge—as valid as physical knowledge was then thought to be—in the field of living forms and above all in the field of human conduct.[5]

Pearson was obviously influenced by *Natural Inheritance*. His first lecture on inheritance was given shortly after its publication and consisted of an exposition and amplification

4. Quoted in Egon Sharpe Pearson, *Karl Pearson* (Cambridge: Cambridge University Press, 1938), p. 7.
5. Pearson, in *Speeches,* pp. 22–23.

of Galton's views.[6] At this time he was editing the second
volume of Todhunter's *History of the Theory of Elasticity*
(published 1893) and formulating the views of methodology
in his influential *Grammar of Science* (published 1892). But
Pearson's interest in evolution, heredity, and statistics was becoming stronger. When the biologist W. F. R. Weldon was
appointed to University College in 1891, he exerted a strong
influence on Pearson, whose work was then redirected toward a furtherance of Galton's efforts. Weldon was looking
for someone like Pearson to help him.

WELDON, PEARSON, AND BIOMETRY

Walter Frank Raphael Weldon was born in 1860. He studied botany and zoology one year at the University of London
with Daniel Oliver, Ray Lankester, and A. H. Garrod, in-
tending to enter the medical profession. In 1878 he went to
St. John's College, Cambridge, and began to study with the
young morphologist, Francis Balfour, who greatly influenced
him.

Following the lead of von Baer and Haeckel, Balfour believed that the development of the individual recapitulated
the history of the species and that evolutionary relationships
were often best revealed by a comparative study of embryological development rather than of adult stages. Balfour's
ability was recognized early. He was elected to the Royal
Society at age twenty-seven and published his influential
Comparative Embryology in 1881. Balfour was concerned
with elucidating the relationships between groups of animals, especially those which lay in the amorphous region between the vertebrates and invertebrates.

The excitement of pursuing Darwin's ideas into the embryological realm was a great inducement to all of Balfour's
students, and Weldon became eager to follow in his steps.
He was even given the privilege of working as demonstrator

6. Karl Pearson, "Walter Frank Raphael Weldon, 1860–1906," *Biometrika,* 5 (1906): 16, n.

for Balfour one term. Unfortunately, Balfour was killed in an Alpine accident in 1882, at age thirty-one.

Adam Sedgwick, formerly Balfour's demonstrator, was appointed to Balfour's chair and invited Weldon to demonstrate for him. Weldon soon finished his first published paper, on the early development of *Lacerta muralis,* a lizard. Several other papers on embryology followed, and in 1884 he was appointed University Lecturer in Invertebrate Morphology at St. John's. Many of Weldon's students became biologists. Among them was William Bateson.

Beginning in 1888 Weldon's interest began to turn from morphology to problems in variation and organic correlation. For example, he had observed that evolutionary changes in adults of some species were accompanied by changes in the larval forms; yet the new adult characters and the new larval characters had no apparent connection. Weldon suspected a correlation existed but did not know how to prove it. Although he was an accomplished morphologist, he became convinced that the analysis of evolution by strictly morphological methods was inadequate.

In 1889, Galton's *Natural Inheritance* furnished Weldon what he was seeking—a quantitative method of attacking organic correlation and the problems of variation. He immediately set to work with elaborate measurements of Decapod Crustacea and found the distribution of variations very similar to that found by Galton in man. His paper, entitled "The Variations Occurring in Certain Decapod Crustacea: 1. *Crangon vulgaris,*"[7] was submitted to the Royal Society with Galton as referee. Galton encouraged Weldon and helped him revise the rather primitive statistical treatment. This marked the beginning of a long friendship.

In his next paper, "On Certain Correlated Variations in *Crangon vulgaris,*"[8] Weldon attempted to measure numerically the amount of interrelation between characters in the same individual, that is, the correlation coefficient. He believed that the correlation coefficient between two organs or

7. *Proceedings of the Royal Society* 47 (1890): 445–53.
8. Ibid., 51 (1892): 2–21.

characters would be constant for a given species (or at least races of species) and would clarify the "functional correlations between various organs which have led to the establishment of the great sub-divisions of the animal kingdom." [9] In other words, the evolutionary relationships which traditional morphology had attempted to demonstrate might be better demonstrated by appropriate statistical studies of populations. Weldon stated in a third paper:

It cannot be too strongly urged that the problem of animal evolution is essentially a statistical problem: that before we can properly estimate the changes at present going on in a race or species we must know accurately (a) the percentage of animals which exhibit a given amount of abnormality with regard to a particular character; (b) the degree of abnormality of other organs which accompanies a given abnormality of one; (c) the difference between the death rate per cent in animals of different degrees of abnormality with respect to any organ; (d) the abnormality of offspring in terms of the abnormality of parents and *vice versa*. These are all questions of arithmetic; and when we know the numerical answers to these questions for a number of species, we shall know the deviation and the rate of change in these species at the present day—a knowledge which is the only legitimate basis for speculations as to their past history, and future fate. [10]

With these words Weldon formulated the basic principles of the biometrical approach derived from Galton. He did not, however, know enough mathematics to develop the needed methods, so he began a study of French mathematicians who wrote on probability and attempted to interest a mathematician in his work. From his studies Weldon became an adequate but certainly not brilliant statistician. Far more important, he attracted the attention of Karl Pearson, who developed the basic methods for the statistical study of populations.

Weldon came to University College, where Pearson was teaching, in 1891. Pearson describes their animated conversations:

9. Ibid., p. 11.
10. "On Certain Correlated Variations in *Carcinus moenas*," *Proceedings of the Royal Society* 54 (1893): 329.

[we] both lectured from 1 to 2, and the lunch table, between 12 and 1, was the scene of many a friendly battle, the time when problems were suggested, solutions brought, and even worked out on the back of the menu or by aid of pellets of bread. Weldon, always luminous, full of suggestions, teeming with vigor and apparent health, gave such an impression to the onlookers of the urgency and importance of his topic that he was rarely, if ever, reprimanded for talking "shop."[11]

Weldon's enthusiasm was contagious and Pearson became very interested in the problems of evolution. When he had finished work on the *Grammar of Science* and the second volume of Todhunter's *History of the Theory of Elasticity,* Pearson began to devote much of his thought to evolution. His first paper[12] came as a response to a problem uncovered by Weldon, who had found that the relative frontal breadth in the shore crab did not follow a Gaussian distribution whereas the distributions of other characters of the crab were normal. Weldon hypothesized, and Pearson showed mathematically, that relative frontal breadth must be dimorphic, each form representing a race. In this paper Pearson developed the method of moments for fitting a theoretical curve to observational data; in later years his methods for doing this became more sophisticated.

By late 1893, Weldon and Galton had become good friends. Galton was impressed by Weldon's early papers, and both were interested in discovering other dimorphic characters. Weldon had also begun experiments attempting to measure the selective death rate in several different species. In December of 1893, Galton and Weldon, with several others, worked out a proposal to the Royal Society for the establishment of a committee to further their work. Weldon naturally assumed that the committee would provide some funds and facilities for research, as well as a convenient means of publication. The committee was approved with Francis Galton as chairman and Weldon as secretary. Francis Darwin, A. Macalister, R. Medola, and E. B. Poulton were the other members. The committee was entitled Committee for Conducting Statistical

11. Pearson, "Weldon," p. 18.

12. "Contributions to the Mathematical Theory of Evolution," *Philosophical Transactions of the Royal Society* 185, A (1894): 70–110.

Inquiries into the Measurable Characteristics of Plants and Animals, and it was suggested that a statistician should be added later.

The committee at first seemed to be a great boon for Weldon. He believed the possibilities for extending his researches were boundless, and he set to work with great enthusiasm. Pearson stated that Weldon at this time "wanted the whole mathematical theory of selection, the due allowances for time and growth, the treatment of selective death-rates and the tests of heterogeneity and dimorphism settled in an afternoon's sitting." [13] But as the committee's work progressed, Weldon found the situation far less conducive to research than he had first imagined.

Weldon was unable to distinguish the possible tasks from the impossible, and he felt acutely the lack of powerful mathematical methods, developed only later by Pearson. For example, Weldon attempted to distinguish the subraces of ox-eyed daisies by examining the ray florets. When the material produced strikingly irregular frequency distributions, Weldon was at a complete loss to analyze the data. Also, typical of any committee of the Royal Society, pressure existed to produce solid work rather than tentative conclusions. Weldon therefore had to give up plans for some of his projected researches.

A more basic problem was a disagreement between Galton and Weldon on whether evolution was continuous or discontinuous. As described above, Galton firmly believed that evolution proceeded by jumps, and he expressed this view clearly in 1894 as well as earlier. On the other hand, Weldon and Pearson, both confirmed Darwinists, believed that evolution proceeded by selection operating upon continuous differences. Pearson expressed his belief in the essential gradualness of evolution as early as 1883,[14] and Weldon appears to have become fixed in this belief by the time of his first statistical paper in 1890. The statistical methods used by Pearson and Weldon were particularly suited to the study of continuous variation.

Here then is the paradox. Pearson and Weldon, viewing

13. Pearson, "Weldon," p. 24.
14. E. S. Pearson, *Karl Pearson,* p. 13.

Galton as the founder of the methods of the biometrical school, believed they were following in his footsteps. But they also believed in continuous evolution, while Galton's reasoning, in Pearson's words, "left him practically in the ranks of the mutationists—a strangely inconsistent position for one who has been looked upon as the founder of the Biometric School!" [15]

How did Pearson and Weldon justify their position? Galton's reason for believing in discontinuous evolution has been stated already: the force of regression was so powerful that selection of continuous variations could have only a limited effect; therefore, evolution must proceed by large stable jumps. Pearson and Weldon claimed that Galton simply misinterpreted his own valid methods. Regression did not quash all exceptional variation of the blending sort, *if the exceptional offspring were bred among themselves*. Pearson stated:

The flaw in Galton's argument is . . . that he is overlooking the fact that he has clubbed together parents of all possible types of ancestry, and the "regression" of his sons is solely due to the large number of such parents who have sprung from an ancestry mediocre or below mediocrity. The amount of filial regression depends entirely on the amount of this mediocrity, and there will be no regression if two or three generations above the parents are of like deviation from mediocrity. Thus, although it may still be a matter for experiment and discussion, whether evolution proceeds by variations proper or by spurts, whether it be continuous or advance by jerks, the reason which made Galton the pioneer in advocating discontinuous evolution was a misinterpretation of his own discovery of "regression." [16]

Thus Pearson believed that he and Weldon were following the true Galtonian methods in dealing with evolution, while Galton himself was led astray by bad reasoning. Wilhelm Johannsen later produced new evidence which appeared to contradict Pearson's reasoning and claimed that the true Galtonian method necessarily led to a belief in the discontinuity of evolution (see chap. 4).

15. Pearson, *Galton*, 3A:86.
16. Ibid., p. 79.

Galton differed markedly from Weldon on the interpretation of the process of evolution. Both wanted the committee to study dimorphic forms, but Galton did not think one form could be continuously selected into another and Weldon did. Galton saw each of the dimorphic forms as a stable center which resisted the influence of selection. Weldon did not see any significant obstacle to selection. The outcome of this difference was not a personal quarrel but could be seen in the way Galton shaped the aims of the committee.

In 1895, Weldon published a paper which formed part of the first report of the committee. It was entitled "Attempt to Measure the Death-rate Due to the Selective Destruction of *Carcinus moenas* with Respect to Particular Dimension."[17] Weldon tried to demonstrate that the death rate was correlated with a measurable character of the shore crab. If this were true, the Darwinian theory of gradual evolution by the selection of continuous differences would be demonstrated.

Galton could have hardly agreed with Weldon's conclusions. He needed to say nothing, however, because William Bateson entered vociferously into a sharp criticism of Weldon's methods and conclusions. Bateson considered it a disgrace that Weldon should be allowed to publish such papers under the auspices of a committee of the Royal Society and made his views known to Galton in a series of long letters. From this time on Bateson was inextricably involved with the committee and, more than any other individual, shaped its future work.

WILLIAM BATESON AND DISCONTINUOUS EVOLUTION

William Bateson was born in 1861. His father, Dr. William Henry Bateson, was forty-nine years old at the time and master of St. John's College, Cambridge, a post he filled until his death in 1881. Young William was not very happy and did poorly in school. At age fourteen he went to Rugby, a preparatory school. Later, Bateson's wife had this to say:

But in spite of his very evident ability, Will was no success at school. Quarter after quarter his school reports express the

17. *Proceedings of the Royal Society* 57 (1895): 360–79.

dissatisfaction and disappointment of his masters, and his name figures ominously near the bottom of all his class lists. He was unpopular among the boys. Probably his intense and emotional sensitiveness, combined with an unusually alert critical faculty, made him an object of dislike to his school-fellows, and made his masters objects of dislike to him.[18]

Bateson himself wrote his mother during his stay at Rugby:

Is anyone happy? I don't think I shall be. You will say, this is all morbid nonsense, but it is true. I never get on with anybody for long; at home even I am always in some scrape except when I am alone. And don't please write back that I am foolish and that, and then not tell me how to cure it.[19]

Even when Bateson was successful in his field, he retained his sensitivity to criticism and responded quickly and sharply to his critics.

In 1879, Bateson left Rugby and entered St. John's College, where his father was still master. The mathematics portion of the elementary matriculation exam gave him trouble:

Mathematics were my difficulty. Being destined for Cambridge, I was specially coached in mathematics at school [Rugby]. Arrived here [St. John's], I was again coached, but failed. Coached once more I passed, having wasted, not one, but several hundred hours on that study.[20]

Bateson never became competent in mathematics—a sore point in his later controversy with the biometricians.

At St. John's, however, Bateson was a successful student. He graduated in 1883 after placing first in both parts of the Natural Sciences Tripos. W. F. R. Weldon, who entered one year before Bateson, was at St. John's at this time studying embryology. Mrs. Bateson states that in 1883 Weldon was Bateson's "most intimate friend," [21] and Weldon was instrumental in getting Bateson interested in the wormlike *Balanoglossus*. Weldon not only gave Bateson access to his own col-

18. Beatrice Bateson, *William Bateson, F.R.S. Naturalist* (Cambridge: Cambridge University Press, 1928), pp. 4–5.
    19. Ibid., p. 5.
    20. Ibid., p. 10.
    21. Ibid., p. 17, n.

lections but also helped him get permission to study with Professor W. K. Brooks of Johns Hopkins University during the summers of 1883 and 1884. *Balanoglossus* was abundant in Chesapeake Bay. Despite the friendship, a sour note prophetic of the future was later revealed by Bateson, who said he was "often made to feel like Weldon's bottle-washer"[22] during his student days.

Bateson's careful study of *Balanoglossus,* an animal which had been previously classified as an Echinoderm, was his only research in traditional morphology. He published three descriptive papers; then in a fourth entitled "The Ancestry of the Chordata,"[23] he discussed the significance of his work. In a boldly conceived argument, Bateson showed that segmentation, which *Balanoglossus* lacked, was not a basic characteristic of the chordates, and that in other respects the animal should be considered a primitive member of Chordata. He then elucidated the relationship of *Balanoglossus* and its allies to the other chordates. His argument was a classic example of the application of Balfour's embryological method and was widely incorporated into textbooks.

But Bateson was already growing beyond Balfour's method. Even as he wrote the paper on the ancestry of the Chordata, he stated his reservations:

The decision that it would be profitable to analyse the bearing of the new fact in the light of modern methods of morphological criticism, does not in any way prejudge the question as to the possible or even probable error in these methods.

Of late the attempt to arrange genealogical trees involving hypothetical groups has come to be the subject of some ridicule, perhaps deserved. But since this is what modern morphological criticism in great measure aims at doing, it cannot be altogether profitless to follow this method to its logical conclusions.

22. Recorded by R. C. Punnett, "Early Days of Genetics," *Heredity* 4 (1950): 2.

23. *Quarterly Journal of Microscopical Science* 26 (1886); reprinted in R. C. Punnett, ed., *Scientific Papers of William Bateson,* 2 vols. (Cambridge: Cambridge University Press, 1928), 1:1–31 (hereafter cited as *Scientific Papers*).

That the results of such criticism must be highly specula-
tive, and often liable to grave error, is evident.[24]

Mrs. Bateson notes that within two years her husband "out-
grew the *Balanoglossus* work and came even to regard it as
trifling." [25]

Bateson, as did Weldon when he became disenchanted
with Balfour's work, turned to the study of variation as the
key to the unsolved problems of evolution. It was probably
Brooks who guided Bateson's interests in this direction. During
the summers of 1883 and 1884, Bateson and Brooks engaged
in long conversations about variation and the mechanism of
evolution. Mrs. Bateson said that her husband "delighted in
recalling the long hours of discussion (Brooks lying in his
shirt-sleeves on his bed and Will sitting by), when problem
and theory and practice passed in long review with ever fresh
interest." [26] In 1883, Brooks was just finishing his book *The
Law of Heredity: A Study of the Cause of Variation and the
Origin of Living Organisms,* in which he proposed a new
theory of heredity to supplant Darwin's pangenesis. His theory of heredity allowed for saltation variation and discontinuous evolution. In the section entitled "Saltatory Evolution,"
Brooks cites the arguments of Huxley, Galton, and Mivart
concerning saltation evolution, then gives a series of examples
of new races being formed by sudden jumps. He concludes:

These cases show us that very considerable variations may
suddenly appear in cultivated plants and domesticated animals, and that these sudden modifications may be strongly
inherited, and may thus give rise to new races by sudden
jumps.
The analogy of domesticated forms would lead us to believe that the same thing sometimes occurs in nature, and
that Darwin has over-estimated the minuteness of the changes
in wild organisms, and has thus failed to see that natural
selection may give rise to new and well-marked races in a few
generations.[27]

24. Ibid., p. 1.
25. B. Bateson, *William Bateson,* p. 18.
26. Ibid.
27. W. K. Brooks, *The Law of Heredity: A Study of the Cause of
Variation and the Origin of Living Organisms* (Baltimore: John
Murphy, 1883), pp. 301–2.

In another section Brooks treated the significance of serial homology and symmetry. The issues raised by Brooks—discontinuity in evolution, symmetry in organisms, and heredity —became the major problems upon which Bateson focused his life work.

Immediately upon finishing his paper on Chordata in the spring of 1886, Bateson left for Russia to investigate the relation between the variations of animals and their environments. His method was to choose environments which differed clearly in some measurable characteristic and to see if variations were correlated with the differences in conditions. The small isolated lakes of different salinity on the Russian steppes seemed ideal sites to test such correlations. He found no general rule: some animals had characters which were uniformly affected by a change in conditions, whereas other animals were entirely unaffected. Thus the evolutionary effects of a change in conditions did not seem as clear to Bateson as it had to Darwin.

Furthermore, in the best example of an animal which did show a correlated alteration with a change in conditions, Bateson thought the animal might revert to its original form if put back into the original environment:

Upon this point I have no evidence; but that the animals would, if they lived and propagated, ultimately regain their former structure appears probable; for, since it can be shown that certain variations are constantly produced by water of certain constitution, it practically follows that the maintenance of these variations depends also on the same cause.[28]

If this were the case, the correlated variations would have negligible evolutionary significance—no permanent changes could be effected. Bateson found little on his trip to indicate that the natural selection of Darwin's "individual differences" had produced new and permanent species.

Bateson regarded his expedition to Russia, and a later

28. William Bateson, "On Some Variations of *Cardium edule* Apparently Correlated to the Conditions of Life," *Philosophical Transactions of the Royal Society*, B, 180 (1889), reprinted in *Scientific Papers*, 1:34.

shorter one to Egypt, as failures.[29] He had discovered no definite connection between the environment and correlated variations. But his failure stimulated him to search for better information about variation. On the basis of his morphological work, Bateson was elected to the Balfour Studentship in November 1887. Ironically, by the time the studentship was awarded to him, Bateson had become thoroughly disenchanted with Balfour's morphological approach, and he used the Studentship to attack the problem of evolution by the study of variation.

Bateson was appalled by the lack of information about variations of plants and animals. He rightly believed that modern researchers had scarcely moved beyond Darwin's work in this field, and he set out to remedy the situation. At first he simply wanted to gather all the data on variation that he could and publish it. Each person could then draw his own conclusions about the mechanism of evolution from the data.

Since the Balfour Studentship provided few funds for research, Bateson applied in 1890 for the Linacre Professorship in Comparative Anatomy at Oxford. He knew that the position would almost certainly be offered to Ray Lankester, but in case Lankester should refuse, Bateson wanted to offer his new approach to the problems of evolution. His letter of application states clearly his aims at the time. He rejected anew the embryological method of von Baer and Balfour which he said "rests on an error in formal logic." [30] That ontogeny reproduces phylogeny was not a valid assumption. Instead, Bateson declared, variation was the key to evolution. The letter of application made Bateson appear a Darwinian. Indeed, he said the purpose of his research was to "pursue Darwin's problems and to employ Darwin's methods." [31]

29. Mrs. Bateson states that her husband "always regarded these expeditions as failures and regretted that in his inexperience he had undertaken the investigation with too definite and narrow expectation, and had pursued the inquiry too closely to profit by the large opportunity of general observation" (*William Bateson,* p. 27).

30. Ibid., p. 32.

31. Ibid., p. 35.

But already Bateson had become dissatisfied with the study of continuous variations, Darwin's "individual differences." His teacher and friend Brooks had suggested the importance of discontinuous variation. His travels and further research had indicated that evolutionary changes were not directly connected to selection pressures caused by differences in environment acting upon continuous variations. Thus his interest was led inexorably toward the larger, discontinuous variations. By the time of his application for the Linacre Professorship, Bateson was already convinced that in repeated parts, such as fingers or teeth, large variations played a crucial role in evolution. He knew this was distinctly un-Darwinian, but in his application merely said that the importance of the facts he had collected about variation of repeated parts "lies in their value as evidence of the magnitude of the integral steps by which variation proceeds. . . ." [32] Bateson did not want his application to appear un-Darwinian.

As expected, Lankester was appointed Linacre Professor. Bateson felt less restraint to conceal his attitude and began publishing a series of papers on large discontinuous variations. In the first paper, still hesitant to make the break with Darwinism, he declined the rather obvious temptation to draw inferences from the data to the mechanism of evolution: "Though one is naturally tempted to draw seemingly obvious deductions from the facts about to be given, it is not proposed on the present occasion to do more than describe the actual structures as they are found." [33] In the next paper, however, entitled "On the Variations in Floral Symmetry of Certain Plants Having Irregular Corollas," [34] the break is stated clearly.

In the Introduction, Bateson made it clear that he believed Darwin's theory to be impossible. "It is difficult," he said, "to suppose *both* that the process of Variation has been a con-

32. Ibid., p. 36.
33. William Bateson, "On Some Cases of Abnormal Repetition of Parts of Animals," *Proceedings of the Zoological Society* (1890); reprinted in *Scientific Papers*, 1:113.
34. *Journal of the Linnaean Society* (Bot.), 28 (1891); reprinted in *Scientific Papers*, 1:126–61.

tinuous one, and also that Natural Selection has been the chief agent in building up the mechanisms of things." [35] In the shaping of a new character, the small variations which Darwin postulated would be of such small, if any, selective value that natural selection would be ineffective. This was, of course, an old criticism of Darwin's theory, but Bateson believed it to be a real one.

The primary point of the article was to show that discontinuous variations did in fact exist, and in consequence

that in proportion as the process of Evolution shall be found to be discontinuous the necessity for supposing each structure to have been gradually modelled under the influence of Natural Selection is lessened, and a way is suggested by which it may be found possible to escape from one cardinal difficulty in the comprehension of Evolution by Natural Selection.[36]

Bateson concluded that the facts of discontinuous variation presented in the article, while few, "are a sample of the kind of fact which is required to enable us to deal with the problems of Descent." [37]

The moment Bateson broke from Darwinism he also broke from Weldon. Their training in biology had been similar, both being strongly influenced by Balfour. Both rejected the morphological approach and began a study of variation as the key to evolution. But Weldon stayed with Darwin's view of evolution by natural selection of small differences, while Bateson, disillusioned by this approach, adopted the view that evolution proceeded by discontinuous leaps. Bateson was certainly aware of the break he was making with tradition, and once the decision was made, he defended his position with alacrity.

Bateson found support in the ideas of Francis Galton. He had been corresponding with Galton since the publication of *Natural Inheritance* and had sent Galton offprints of his papers. Bateson later said of Galton:

35. Ibid., p. 128.
36. Ibid.
37. Ibid., p. 150.

The novelty of his thoughts and the freshness of his outlook on nature are not to be found in any other living writer, so far as I know. I often remember the thrill of pleasure with which I first read *Hereditary Genius* and the earlier chapters of *Natural Inheritance*.[38]

In the article on floral symmetry where Bateson made his break with Darwinism clear, the similarity of his ideas to those expressed by Galton in *Natural Inheritance* is striking. Bateson states there are two classes of variation: the continuous, exemplified by the variations studied by Galton in man and by Weldon in shrimp, and the discontinuous, some examples of which he had just presented. The distinction Bateson makes between continuous and discontinuous variation is precisely the distinction Galton makes between "sports" and "variation proper." Bateson also uses Galton's notion of equilibrium in making the distinction. The intermediate forms between two discontinuous variations, where symmetry was involved, were "points of unstable equilibrium." [39] Bateson concluded with a statement which sounded like Galton:

If . . . as may be alleged, there is little evidence that species may arise by what may be called discontinuous Variation— a Variation in kind—there is still less evidence that new forms can arise by those Variations in degree which at any given moment are capable of being arranged in a curve of Error, and no one as yet has ever indicated the way by which such Variations could lead to the constitution of new forms, at all events under the sole guidance of Natural Selection.[40]

There can be little doubt that Bateson was influenced by his reading of *Natural Inheritance*.

In 1894, Bateson published his huge *Materials for the Study of Variation, Treated with Especial Regard to Discontinuity in the Origin of Species*.[41] He presented 886 cases of discontinuous variation and expanded his views on discontinuity in

38. Bateson to Miss Evelyn Biggs (great-niece of Francis Galton), 7 July 1909, in Pearson, *Galton,* 3A:288.
39. Bateson, *Scientific Papers,* 1:158.
40. Ibid., p. 159.
41. London: Macmillan, 1894.

evolution. Galton gave the book an enthusiastic welcome. After stating that he himself had propounded similar views for many years, which had gone unheard, Galton continued:

> It was, therefore, with the utmost pleasure that I read Mr. Bateson's work bearing the happy phrase in its title of "discontinuous variation," and rich with many original remarks and not a few trenchant expressions.[42]

Bateson also sent a copy of the book to Huxley, who wrote back:

> I see you are inclined to advocate the possibility of considerable "saltus" on the part of Dame Nature in her variations. I always took the same view, much to Mr. Darwin's disgust, and we used often to debate it.[43]

Both Galton and Huxley clearly approved of Bateson's emphasis upon discontinuity in evolution.

Weldon's response to *Materials* was less enthusiastic. He did not believe evolution was discontinuous. He questioned Bateson's claim that a discontinuous variation was a new center of organic stability and not subject to regression. Galton had, of course, made the same claim earlier. Weldon stated in a letter to Bateson:

> About "regression," I will say only this, that Galton was himself a good deal mixed, at least in his exposition, when he wrote *Natural Inheritance:* and that I cannot conceive that characters "which do not mix" are thereby rendered independent of the phenomenon of regression.[44]

Although Weldon and Bateson disagreed on the issue of continuity in evolution, their correspondence at this time, early 1894, shows nothing of the personal antagonisms which would soon be evident.

On 10 May 1894, Weldon's review of *Materials* appeared in *Nature*. First he praised the book:

42. Galton, "Discontinuity in Evolution," p. 369.
43. T. H. Huxley to Bateson, 20 February 1894, in L. Huxley, *Life and Letters of Thomas Henry Huxley,* 2:394.
44. Weldon to Bateson, 15 February 1894, Bateson Papers, Baltimore, no. 13. A collection of Bateson papers is on microfilm at the American Philosophical Society, Philadelphia. The society furnishes a guide to the number system. The Bateson Papers are cited hereafter as BPB.

The whole work must be carefully read by every serious student; there can be no question of its great and permanent value, as a contribution to our knowledge of a particular class of variations, and as a stimulus to further work in a department of knowledge which is too much neglected.[45]

But then he launched into a sharp criticism of Bateson's interpretation of his data, especially his emphasis upon discontinuity in evolution. He challenged Bateson's contention that species were more discontinuous than the environments which produced them and attacked his treatment of discontinuous variation. The resulting impression was that Bateson had done well to study the problem of variation as connected with evolution but that his ideas of discontinuity in variation and evolution were misguided, as was his method of research. Weldon suggested Bateson should drop his idea of discontinuity and adopt biometrical methods for the study of variation.

Pearson's interpretation was that this review signaled the beginning of Bateson's attacks upon Weldon.[46] Certainly Bateson was provoked by the review. To make matters worse, *Materials* sold poorly. It ran against the grain of Darwinism, which was popular in England at this time. Galton had already received the cold shoulder with his ideas on discontinuous evolution. But Bateson was eager to challenge the orthodox school, and a series of confrontations began between Bateson and the Darwinians, especially Weldon. Within a year of Weldon's review, he and Bateson were engaged in a heated public controversy. They were to cease fighting only with Weldon's death in 1906.

The Public Controversies

THE CINERARIA CONTROVERSY

The first public controversy between Bateson and the Darwinists was over the origin of cultivated Cineraria. At a meeting of the Royal Society on 28 February 1895, the biologist W. T. Thiselton-Dyer exhibited two forms of Cineraria: the wild type *C. cruenta* from the Canary Islands and a re-

45. W. F. R. Weldon, "The Study of Animal Variation," *Nature* 50 (1894): 24.
46. Pearson, *Galton,* 3A:287.

cently cultivated form from the Royal Gardens at Kew. The two varieties differed markedly in the shape and color of the flowers. Dyer claimed that the cultivated form was derived from the wild type by artificial selection of continuous differences. In a letter to *Nature* published on 14 March 1895, Dyer clarified his remarks at the meeting. He minimized the value of sports in evolution: "As I conceive the [evolutionary] process, it is one of continuous adjustment of 'slight' variations on one side and the other." [47] As an example of a change which had been accomplished by the gradual accumulation of small variations, he referred to the change from *C. cruenta* to the modern Cineraria.

Bateson did not attend the meeting but he read Dyer's letter. Here was his chance to challenge the prevalent Darwinian view of evolution. After making a study of horticultural records, he concluded in a letter published in *Nature:*

The foregoing notes of history must, I think, be taken to show (1) that the modern Cinerarias arose as hybrids derived from several very distinct species; (2) that the hybrid seedlings were from the first highly variable; (3) that 'sports' of an extreme kind appeared after hybridization in the early years of the 'improvement' of these plants; (4) that the subsequent perfection of the form, size and habit has proceeded by a slow process of selection. Mr. Dyer's statement that the modern Cinerarias have been evolved from the wild *C. cruenta* 'by the gradual accumulation of small variations' is therefore, in my judgment, misleading, for this statement neglects two chief factors in the evolution of the Cineraria, namely, hybridization and subsequent 'sporting.' [48]

The controversy now began in earnest. Ten additional letters concerning Cineraria were published by *Nature* during the next two months. First, Dyer and Bateson again exchanged letters. Dyer [49] challenged Bateson's belief that hybridization was important in the evolution of the new form of Cineraria because it, like *C. cruenta,* was herbaceous. The other Cinerarias which Bateson claimed were hybridized with

47. *Nature* 51 (1895): 461.

48. Letter from William Bateson, 25 April 1895, ibid., p. 607.

49. Letter from W. T. Thiselton-Dyer, 29 April 1895, *Nature* 52 (1895): 3–4.

C. cruenta to form modern Cineraria were shrubby species. Since modern Cineraria was herbaceous and had leaves like *C. cruenta,* Dyer claimed that modern Cineraria arose directly from *C. cruenta.* Bateson replied that Dyer had not disproved his contention that the change was discontinuous, and that the question of the hybrid origin of cultivated Cineraria was "of subordinate interest" [50] compared to the question of discontinuity. Furthermore, Dyer's claim against the hybrid origin of modern Cineraria was, said Bateson, directly contradicted by the horticultural literature.

Weldon now entered the fray with an attempt to discredit Bateson's position. His argument was that Bateson had misused his source materials: "I . . . wish to point out that Mr. Bateson has omitted from his account of these records some passages which materially weaken his case. . . . All I wish to show is that the documents relied upon by Mr. Bateson do not demonstrate the correctness of his views; and that his emphatic statements are simply of want of care in consulting and quoting the authorities referred to." [51]

Bateson thought Weldon had initiated a personal attack. They arranged to meet on 21 May 1895 to discuss the issue. During the conversation Bateson understood Weldon to say that Dyer was bluffing. Bateson's recorded notes of the conversation state: "Weldon's position in writing is therefore that of the accomplice who creates a diversion to help a charlatan. I cannot at all understand his motives, or how he can bring himself to play this part." [52] Weldon's position was made clear in a letter to Bateson three days later:

24 May 1895

Dear Bateson,
I can do no more.

First, you accuse me of attacking your personal character; and when I disclaim this, you charge me with a dishonest defense of some one else.

I have throughout discussed only what appeared to me to be facts, relating to a question of scientific importance.

50. Letter from William Bateson, 9 May 1895, ibid., p. 29.
51. Letter from W. F. R. Weldon, 13 May 1895, ibid., p. 54.
52. BPB 10.

If you insist upon regarding any opposition to your opin-
ions concerning such matters as a personal attack upon yourself, I may regret your attitude, but I can do nothing to change
it.

Yours very truly,
W. F. R. Weldon [53]

The two men were never on friendly terms again.

The controversy continued with little change in the basic
position of the antagonists, but their letters became more
polemical. Dyer's letter of 23 May ends with the statement:
"I think that in the study of evolution we have had enough
and to spare of facile theorizing. I infinitely prefer the sober
method of Prof. Weldon, even if it should run counter to my
own prepossessions, to the barren dialectic of Mr. Bateson." [54]

And Bateson's of 30 May concludes: "The facts I have been
able to collect may have been few, but by a study of the
writings of my antagonists, I have not been able to add
materially to their number." [55]

The public controversy died in June of 1895 when *Nature*
refused to publish anything more on the subject, but the
antagonisms it generated had consequences not dreamed of
by the participants. The argument over Cineraria set the
stage for a continuing confrontation between Bateson and the
Darwinists. Bateson knew he was bucking the tide and des-
perately feared his views would be forced into oblivion. He
reacted vigorously against this possibility by starting breeding
experiments which might reinforce his position, and by constant rebuttal of his critics.

THE STRUGGLE OVER THE EVOLUTION COMMITTEE

In 1895 the first report of Galton's committee was issued.
It consisted of two papers by Weldon. The first, already mentioned, dealt with the correlation of death rates with certain
characters of the shore crab. The second, read at the same
meeting as Dyer's first Cineraria paper, was a short broadside
aimed directly at Bateson's belief in the discontinuity of evolu-

53. Ibid.
54. Letter from W. T. Thiselton-Dyer, 13 May 1895, *Nature* 52
(1895): 79.
55. Letter from William Bateson, 26 May 1895, ibid., p. 104.

tion. Weldon did not deny "the possible effect of occasional 'sports' in exceptional cases" but claimed that natural selection of small variations was sufficient to explain the direction and rate of evolution. He further stated that "the questions raised by the Darwinian hypothesis are purely statistical, and the statistical method is the only one at present obvious by which that hypothesis can be experimentally checked." [56] This was a strong statement considering the small amount of evidence Weldon had collected.

Bateson was disturbed by the report of the committee. He found that Weldon had not measured crabs in the same stage of molting. Since the magnitude of the character measured by Weldon changed after each molt, Bateson thought his results were invalid. He wrote a series of four letters to Galton, as chairman of the committee, explaining his criticism of Weldon's work. He even offered to print his letters for distribution to the members of the committee. Galton passed Bateson's criticism on to Weldon. The three then engaged in a flurry of correspondence concerning the report and the aims of the committee. [57]

Galton reacted to Bateson's criticisms with mixed feelings. He believed Bateson was rather pushy. And Bateson's lack of sympathy for the statistical treatment of biological problems, the express purpose of the committee, was obvious. But concerning the mechanism of evolution, Galton agreed with Bateson, not Weldon. Galton's solution to this dilemma was typically idealistic. He would add to the committee Bateson and other evolutionists whose interests were not primarily biometrical. His hope was that the enlarged committee would work to produce a broadly based view of evolution.

Pearson's belief was that Galton suggested Bateson be added to the committee because he "was so weary of Bateson's incessant letters to the committee." [58] But Galton really thought Bateson had much to offer the committee, a view

56. W. F. R. Weldon, "Remarks on Variation in Animals and Plants," *Proceedings of the Royal Society* 57 (1895): 380, 381.

57. BPB 10, 13, and 15. An almost complete record of the correspondence is in the Bateson Papers.

58. Pearson to Galton, 14 July 1906, in Pearson, *Galton,* 3A:290.

Weldon and Pearson did not share. In a letter dated 17 November 1896, Galton pleaded with Weldon: "It would in many ways be helpful, if Bateson were made a member of our Committee, but I know you feel that in other ways it might not be advisable. The other members besides yourself hardly do enough." [59]

Pearson has stated that the difference in Galton's and Weldon's views of evolution "by no means caused friction between the Chairman and Secretary of the Committee." [60] This cannot be entirely true, because both Weldon and Pearson were opposed to a widening of the committee to include Bateson and other nonbiometrical evolutionists, and Galton was definitely in favor of such a move. Both Pearson and Weldon sensed that a committee of diverse interests would be bogged down by controversy, instead of providing a well-rounded view of evolution as Galton hoped.

In December 1896, Pearson joined the committee. On 1 January 1897, Galton wrote Bateson: "We are going to have a Committee meeting soon. Both Weldon and myself are desirous that you should join us. Would it be agreeable to you that we should propose your name?" [61] Bateson replied:

I very much appreciate the suggestion that you and Weldon so kindly make, that I should join the Measurements Committee. On the whole however I think I had better not. I am not convinced that the present lines of inquiry of the Committee are fruitful and I do not think it is likely that the results will be at all proportionate to the labour expended. [62]

But within the month Galton convinced Bateson that the committee would change in a suitable direction. In late January, Bateson, along with F. D. Godman, Ray Lankester, Thiselton-Dyer, and five others joined the committee. At the next meeting on 11 February 1897, with the new members present, it was decided to change the name of the commitee to

59. Galton to Weldon, 17 November 1896, ibid., p. 127.
60. Pearson, *Galton*, 3A:126.
61. BPB 15. Weldon was not "desirous" that Bateson should join the committee. This was Galton's way of trying to make Bateson feel welcome.
62. Bateson to Galton, 3 January 1897, ibid.

the Evolution Committee of the Royal Society. Added to its original statement of purpose was the "accurate investigation of Variation, Heredity, Selection, and other phenomena relating to Evolution."

The fears of Weldon and Pearson were immediately realized. The new members showed little sympathy to the biometrical approach, and in some cases, much antagonism. The day after the 11 February meeting Pearson wrote to Galton:

> The Committee you have got together is entirely unsuited. . . . It is far too large, contains far too many of the old biological type, and is far too unconscious of the fact that the solutions to these problems are in the first place statistical, and in the second place statistical, and only in the third place biological.[63]

Of course many of the new members found this attitude antagonistic and for the next three years the committee was largely disorganized.

During these three years Galton and Pearson published several papers on the law of ancestral heredity.[64] These papers left a trail of confusion about the meaning and application of the law.[65] They also contributed to Bateson's disillusionment with the biometrical approach to the problem of evolution and increased his desire to redirect the aims of the Evolution Committee.

In *Natural Inheritance* Galton had stated his law tentatively because he admittedly had insufficient evidence. At that time, 1889, he hesitated to apply it beyond the grandparental generation. But in 1897, with new data in hand on the inheritance of coat color in Basset hounds,[66] Galton was emboldened to state his law as follows, in what we shall term form A:

> The two parents contribute between them on the average one-half, or (0.5) of the total heritage of the offspring; the

63. Pearson to Galton, 12 February 1897, in Pearson, *Galton,* 3A:128.
64. Note that Pearson did not give the law this name until 1898.
65. Much of this confusion was caused by technical considerations. What follows in the text is only a summary of these considerations. For a full exposition and support for statements made in the text, see the Appendix. The text should be read before the Appendix.
66. Galton's data came from records kept by Sir Everett Millais over a period of twenty years.

four grandparents, one-quarter, or $(0.5)^2$; the eight great-grandparents, one-eighth, or $(0.5)^3$, and so on. Thus the sum of the ancestral contributions is expressed by the series $(0.5) + (0.5)^2 + (0.5)^3$, etc., which, being equal to 1, accounts for the whole heritage.[67]

Below on the same page Galton stated his law in another form. Supposing the deviation of the offspring from the mean M to be D, the deviation of the parents from M to be D_1, the deviation of the grandparents from M to be D_2, etc., Galton claimed his law took the form B:

$$M + D = \tfrac{1}{2}(M + D_1) + \tfrac{1}{4}(M + D_2) + \text{etc.}$$
$$= M + (\tfrac{1}{2}D_1 + \tfrac{1}{4}D_2 + \text{etc.}).$$

Galton said this form of the law showed that "the law may be applied *either* to total values or to deviations."

Unfortunately, the two forms in which Galton stated his law are mathematically inconsistent; yet he used them interchangeably in his calculation. This of course led to much confusion. Forms A and B caused other problems. They were statistical statements of phenotypic resemblances. But Galton claimed form B could be inferred a priori from the physiology of the hereditary process. Thus many biologists were led to believe that if they disproved Galton's conception of the physiology of heredity, a task easily accomplished after the rediscovery of Mendelian inheritance, they also disproved his law of ancestral heredity. As stated, however, forms A and B were statistical statements of phenotypic resemblances and could hold whatever the physiology of heredity might be, as Pearson continually pointed out. One further problem which added to the confusion was Galton's lack of clarity about the sort of variation to which his law applied. It applied to characters which were inherited discontinuously, such as eye color, but not to sports, which were also inherited discontinuously, or to characters involved in hybridizations. It was anything but obvious where Mendelian characters fit into this scheme, and the confusion of the Mendelians after 1900 concerning Galton's law is understandable.

67. Francis Galton, "The Average Contribution of Each Several Ancestor to the Total Heritage of the Offspring," *Proceedings of the Royal Society* 61 (1897): 402.

Galton's formulation of his law in 1897 was confusing enough, but that was just the beginning. The paper on inheritance in Basset hounds stimulated Pearson to reevaluate Galton's law. Pearson wrote a paper in which he revised Galton's law considerably, beyond anything Galton might have done on his own. Then Pearson promptly christened his new creation "Galton's Law of Ancestral Heredity." [68] The resulting confusion of biologists about Galton's law was to make Pearson wish he had used a different choice of words. Pearson later had to point out over and over how his own conception of the law of ancestral heredity differed from that of Galton.

Pearson's expression of the law of ancestral heredity was mathematically far more sophisticated than Galton's and in some ways bore little resemblance to Galton's. It had several extra variables, was based upon statistical correlations not upon resemblances as Galton had imagined, and no longer applied to nonblending inheritance, as in Galton's case with Basset hounds. The Basset hound data had of course supplied Galton with his only empirical "proof" of his law. Furthermore, Pearson showed that his conception of the law was completely consistent with gradual Darwinian selection, whereas Galton had repeatedly expressed the belief that regression and continuous evolution were inconsistent. Pearson concluded his 1898 paper with the following glowing statement about "Mr. Galton's Law" (as revised by Pearson):

It is highly probable that it is the simple descriptive statement which brings into a single focus all the complex lines of hereditary influence. If Darwinian evolution be natural selection combined with *heredity,* then the single statement which embraces the whole field of heredity must prove almost as epoch-making to the biologist as the law of gravitation to the astronomer.[69]

Pearson sent the paper to Galton as a New Year's greeting, 1 January 1898. Galton answered on 4 January: "You have indeed sent me a most cherished New Year greeting. It delights me beyond measure to find that you are harmonizing

68. Karl Pearson, "Mathematical Contributions to the Theory of Evolution. On the Law of Ancestral Heredity," *Proceedings of the Royal Society* 62 (1898): 386.
69. Ibid., p. 412.

what seemed disjointed, and cutting out and replacing the rotten planks of my propositions." [70] Thus Galton, now nearly seventy-six years old, gave his hearty approval to Pearson's revisions of his law of ancestral heredity.

The result was nearly complete confusion surrounding the meaning of the law of ancestral heredity. Not only were Galton's statements of the law mathematically inconsistent and unclear in their relationships to hereditary processes, but also Pearson's revised law, going by the name of Galton's law, was significantly different from anything Galton had previously imagined. Biologists were naturally confused about the meaning and application of Galton's law. Some believed it meant one thing, and some another. Usually they knew that Pearson had revised the law, but they could not follow his mathematics and clung to Galton's more accessible statements of it. Probably the only persons to correctly understand Pearson's revisions of Galton's law were Pearson and some of his students. The confusion surrounding Galton's law was so complete that biologists never straightened it out. The rise of population genetics showed that Galton's law was irrelevant and it simply dropped from sight.

One immediate consequence of Galton's and Pearson's papers in 1897 and 1898 was to convince Bateson, though he could not follow the mathematics, that Pearson had subverted Galton's ideas on heredity from discontinuous evolution to the cause of gradual Darwinian evolution. Moreover, Pearson had apparently captured the aging Galton's favor in this endeavor. Bateson became determined not to lose the Evolution Committee to the work of Pearson, Weldon, and Galton.

In 1897, soon after he joined the committee, Bateson was awarded a small grant which he used to begin experiments in poultry and plant breeding. He hoped to turn the interests of the committee to research of this sort. Pearson and Weldon were annoyed by the whole situation and attempted to disband the committee rather than have it fall into the hands of the opposition. In a letter dated 5 June 1899, Weldon expressed to Bateson his belief that "the Evolution Committee is a mis-

70. Pearson, *Galton,* 3B:504.

take."[71] On 6 November 1899, Weldon was again writing Bateson, this time to arrange a meeting in which the committee would determine its fate.[72] Weldon was clearly sick of the committee and hoped it would disband. Bateson wanted to save the committee as a source of publication and financial support for research on the problems of variation.

Recognizing his failure to create a viable committee, Galton resigned on 25 January 1900. Pearson and Weldon resigned at the same time hoping the Evolution Committee would collapse. But Bateson gained enough support to save it. In February 1900, the committee elected Godman as chairman and Bateson as secretary. The efforts and reports of the Evolution Committee became exclusively devoted to the work of Bateson and his followers.

Galton, Pearson, and Weldon remained close friends. But it was probably Galton's adherence to discontinuity in evolution which led to his widening of the committee against the wishes of Weldon and Pearson, and thus to an increase of hostilities between the biometricians and Bateson's group.

Each side believed its position was under heavy attack. The biometricians were extremely unhappy about the fate of the committee. Pearson later reported bitterly that "a definite plan was formed about 1896 to eject the biometricians and take possession of the Evolution Committee," and that "the capture of the Committee was skilful and entirely successful."[73] What had appeared to be an ideal committee had been subverted to antagonistic research, and an avenue for publication of biometrical work closed. In addition to the attack from the Batesonians, the biometricians were feeling the brunt of the resentment from the old guard Darwinists, who had little appreciation of the new statistical analysis of evolution. Bateson and his followers had consolidated a new position in the Evolution Committee but still felt that position precarious. They were determined to have their views recognized. Thus the situation was already tense at the time Mendel's work on heredity was rediscovered.

71. BPB 15.
72. Ibid.
73. Pearson, *Galton*, 3A:287, 127.

3

The Conflict Between Mendelians and Biometricians

MENDEL'S THEORY OF HEREDITY, REDISCOVERED IN 1900 BY HUGO de Vries, Carl Correns, and Erich von Tschermak, intensified the already heated controversy about the continuity of evolution. After 1900, when Bateson became a champion of Mendelism and Pearson named his science biometry, the controversy became known to the public as the conflict between the Mendelians and biometricians. The conflict drove a wedge between Mendel's theory of heredity and Darwin's theory of continuous evolution and consequently delayed the synthesis of these theories into population genetics.

Bateson began breeding experiments in 1897. He had not discovered Mendelian ratios by 1900, but he was prepared to understand the results of Mendel's experiments. In 1899 he proposed experiments similar to those of Mendel:

What we first require is to know what happens when a variety is crossed with its *nearest allies*. If the result is to have scientific value, it is almost absolutely necessary that the offspring of such crossing should then be examined *statistically*. It must be recorded how many of the offspring resembled each parent and how many shewed characters intermediate between those of the parents. If the parents differ in several characters, the offspring must be examined statistically, and marshalled, as it is called, in respect of each of those characters separately.[1]

Mendel would have liked the proposal.

Bateson expected the results from his experiments to support the theory of discontinuous evolution. He said there were two primary problems with Darwin's idea of natural selection: the selective value of small variations was negligi-

1. William Bateson, "Hybridization and Cross Breeding as a Method of Scientific Investigation," *Journal of the Royal Horticultural Society* 24 (1900); reprinted in B. Bateson, *William Bateson*, p. 166.

ble, and the "swamping effect of intercrossing" obliterated the variation upon which selection acted. He believed that both difficulties disappeared if selection acted upon large discontinuous variations. Such variations had high selective value and were not obliterated by intercrossing.

By 1899, Bateson was prepared to understand Mendel's experiments dealing with discontinuous variations. On 8 May 1900, he was on his way to the Royal Horticultural Society to read a paper entitled "Problems of Heredity as a Subject for Horticultural Investigation." Mrs. Bateson later told the story:

> He had already prepared this paper, but in the train on his way to town to deliver it, he read Mendel's actual paper on peas for the first time. As a lecturer he was always cautious, suggesting rather than affirming his own convictions. So ready was he however for the simple Mendelian law that he at once incorporated it into his lecture.[2]

Bateson was happy with his find. Mrs. Bateson remarked: "Mendel's work fitted in with Will's with extraordinary nicety. . . . His delight and pleasure on his first introduction to Mendel's work were greater than I can describe."[3]

Bateson believed that Mendelian heredity, which treated discontinuous variations and prevented swamping, was the perfect theory to complement the discontinuous theory of evolution. When published, the lecture in which he first mentioned Mendel contained the statement:

> These experiments of Mendel's were carried out on a large scale, his account of them is excellent and complete, and the principles which he was able to deduce from them will certainly play a conspicuous part in all future discussions of evolutionary problems.[4]

Bateson's conclusion was that Mendelian inheritance supported discontinuous evolution.

Yet his was not the only possible conclusion. Mendel utilized only discontinuous characters in his experiments with peas,

2. *William Bateson,* p. 73.
3. Ibid., pp. 70, 73.
4. William Bateson, "Problems of Heredity as a Subject for Horticultural Investigation," *Journal of the Royal Horticultural Society* 25 (1900); reprinted in B. Bateson, *William Bateson,* p. 175.

but he also described an experiment with two varieties of *Phaseolus*. One variety had white flowers and the other purple. When crossed, the hybrids all produced purple flowers. The seeds from the hybrids produced plants with flowers of a series of colors, from purple red to pale violet to white. Mendel concluded that

> even these enigmatical results, however, might probably be explained by the law governing *Pisum* if we might assume that the colour of the flowers and seeds of *Ph. multiflorus* is a combination of two or more entirely independent colours, which individually act like any other constant character in the plant.[5]

Thus Mendel indicated that his theory might account for even a continuous array of variation. A Darwinian need not reject Mendel's theory as unsuitable for explaining continuous variation and, in turn, continuous evolution.

Pearson and Weldon might have argued that Mendelism supported Darwinian evolution. But Bateson made the more obvious connection between Mendelism and discontinuous evolution. In reaction the biometricians viewed Mendelism as a threat. Consequently the six years following the rediscovery of Mendelism witnessed increasingly bitter confrontations between the Mendelians and biometricians. Each confrontation is treated here as a unit, though several were often in progress at one time. The first controversy began in 1900 before Mendelism had become a heated issue.

THE HOMOTYPOSIS CONTROVERSY

Karl Pearson worked on his theory of homotyposis in the summer of 1899. On 6 October 1900, he submitted to the Royal Society an abstract and read it on 15 November, by which time he had completed the entire memoir.[6] The theory of

5. Gregor Mendel, *Experiments in Plant Hybridization* (Cambridge: Harvard University Press, 1958), p. 30. This translation of Mendel's paper was made by the Royal Horticultural Society, with footnotes and commentary by William Bateson.

6. Karl Pearson, "On the Principle of Homotyposis and Its Relation to Heredity, to the Variability of the Individual, and to That of the Race. Part 1. Homotyposis in the Vegetable Kingdom," *Philosophical Transactions of the Royal Society*, A, 197 (1901): 285–379.

homotyposis was Pearson's attempt to simplify the whole problem of heredity. He argued that: (1) an individual organism produces "undifferentiated like organs," such as blood corpuscles, flower petals, tree leaves, or fish scales. Yet these organs are not exactly alike; the "undifferentiated like organs of an individual possess a certain variability, and . . . this variability is somewhat less than that of all like organs in the race";[7] (2) the sperm cells and ova "may each be fairly considered as 'undifferentiated like organs' ";[8] (3) the offspring are fair representatives of the parental germs; (4) therefore the quantitative resemblance between offspring of the same parent should be the same as the quantitative resemblance between undifferentiated like organs in an individual organism.

The undifferentiated like organs Pearson called "homotypes." Homotyposis was "the principle that homotypes are correlated, *i.e.,* that variation within the individual is less than that of the race, or that undifferentiated like organs have a certain degree of resemblance."[9] Pearson argued that heredity was only a special case of homotyposis. Consequently, "when we ascertain the sources of variation in the individual, then we shall have light on the problem of fraternal resemblance."[10] Since the production of variability in offspring was strictly analogous to the production of undifferentiated like organs in the individual, he argued further that one should theoretically expect no more variability in sexually reproducing species than in asexually reproducing species. Pearson adhered to this belief

7. Ibid., p. 287.
8. Ibid., p. 288.
9. Ibid., p. 294. Pearson calculated the correlation between homotypes as follows. He collected a set number of leaves (or whatever character he was investigating), usually 26, from each of about 100 trees. Each leaf was individually classified or measured. Then for the leaves on each tree he took all the possible pairs, or ½ (26 × 25) = 325 pairs. Then he entered each pair in a correlation table using each member of the pair as the "first" leaf, rendering the table symmetrical. Thus the 325 pairs gave 650 entries in the correlation table. Repeating this procedure for the 99 other trees provided 65,000 entries in all in the table. Using the standard procedures Pearson then calculated from the table the correlation between leaves on a single tree.
10. Ibid., p. 291.

for many years. But curiously, he was soon to be among the first to deduce that in sexually reproducing species the genetic recombination predicted by Mendelian heredity could provide vast numbers of genetic variants.

In the paper Pearson produced a theoretical argument, based on dubious assumptions, that fraternal correlation equalled homotypic correlation. And he presented some sixty pages of data from the vegetable kingdom. The data yielded a mean value of homotypic correlation of 0.4570. From other sources Pearson had already obtained a value of 0.4479 for fraternal correlation. Thus homotypic correlation and fraternal correlation were "sensibly equal." Pearson believed he had proved that the variation of undifferentiated like organs in an individual was the same phenomenon as variation between brothers.

If true, Pearson's homotyposis theory would have been a stunning contribution to biology. Homotyposis was perhaps comparable to the generalization "ontogeny reproduces phylogeny" of von Baer, Haeckel, and Balfour, but its biological foundations were just as weak.

Bateson attended the meeting of the Royal Society when Pearson read his abstract. Pearson reported to Galton that Bateson "came to the R.S. at the reading 'and said there was nothing in the paper." [11] Bateson had been appointed as one of the referees who would decide whether the Royal Society should publish Pearson's completed memoir, and he had prepared detailed criticisms. He even told Pearson at the meeting that he had written an unfavorable report.

When Bateson was writing his criticism of Pearson's homotyposis paper he had been acquainted with Mendel's theory of heredity for almost six months. Using Mendel's theory he could have devastated Pearson's theory. Pearson had assumed the sperm cells and ova were undifferentiated like organs. But Mendel believed his experiments showed conclusively that the germ cells must be differentiated. The translation of Mendel's paper annotated by Bateson himself states: "With *Pisum* it was shown by experiment that the hybrids form egg and pollen

11. Pearson, *Galton*, 3A:241.

cells of *different* kinds, and that herein lies the reason of the variability of their offspring." [12] The offspring of the hybrids were differentiated because the germ cells were differentiated. One germ cell was different from another because it had a different combination of differentiating elements. Mendel said "we must further assume that it is only possible for the differentiating elements to liberate themselves from the enforced union when the fertilizing cells are developed." [13] Thus the differentiation which occurred in the development of a single plant was not differentiation of the germ plasm. That occurred only in the production of germ cells. Variation in a single plant was fundamentally different from variation in the offspring of that plant. Clearly Mendel's theory was contradictory to Pearson's homotyposis theory.

Bateson did not use the criticism from Mendel's theory because he did not believe Mendel's "differentiating elements" were material bodies. As early as 1893, Bateson had developed a "vibratory theory of heredity," which did not fit with a materialist view of heredity, and he maintained this theory with some misgivings to the end of his life. It even caused him to reject the chromosome theory of heredity. [14]

In his published criticism of Pearson's homotyposis paper, Bateson indicated his complete agreement with Pearson's belief "that the relationship and likeness between two brothers is an expression of the same phenomenon as the relationship and likeness between two leaves on the same tree, between the scales on a moth's wing, the petals of a flower, and between repeated parts generally." [15] Evidently Bateson misunderstood or rejected what Mendel had said.

Bateson's actual arguments against homotyposis were: (1)

12. Mendel, *Experiments,* p. 35.
13. Ibid., p. 36.
14. An account of the development of Bateson's thought concerning heredity may be found in William Coleman, "Bateson and Chromosomes: Conservative Thought in Science" (unpublished manuscript to appear in *Centaurus*). Coleman is at Johns Hopkins University, Department of History of Science.
15. William Bateson, "Heredity, Differentiation, and Other Conceptions of Biology: A Consideration of Professor Karl Pearson's Paper 'On the Principle of Homotyposis,'" *Proceedings of the Royal Society* 69 (1901); reprinted in Bateson, *Scientific Papers,* 1:404.

no theoretical distinction existed between differentiation and variation in a single individual or population, as Pearson assumed. Therefore Pearson's category "undifferentiated like organs" had no existence in nature; (2) Pearson ignored the importance of "specific" and "normal" variations, which were Bateson's new names for discontinuous and continuous variations. "Specific" variations were important for evolution but "normal" variations were not. Pearson did not recognize this "fact."

Without Bateson's prior approval, his criticism of Pearson's paper was distributed to the other referees before they had received Pearson's completed memoir. Pearson was greatly disturbed by this unusual procedure and communicated his unhappiness to the Royal Society and to Bateson.

The controversy surrounding Pearson's homotyposis paper precipitated an important development in the struggle between the biometricians and Mendelians. Pearson and Weldon became so disenchanted with publication procedures at the Royal Society that they decided to start a new journal. On 16 November 1900, the day after Pearson presented the abstract of his homotyposis paper, Weldon wrote to Pearson: "Do you think it would be too hopelessly expensive to start a journal of some kind?"[16] Pearson suggested the name *Biometrika;* he said "the 'K' was mine (K.P. not C.P.)."[17] In June of 1901 Cambridge University Press agreed to publish the journal and the first issue appeared in October of that year.

When Pearson objected to the procedure adopted by the Royal Society concerning Bateson's criticism, Bateson immediately withdrew his paper until Pearson's was published. He also wrote a letter of apology to Pearson, who responded with a pleasant letter commending Bateson's action. Pearson told Bateson in this letter that the new journal *Biometrika* would not "intend to be exclusive 'Nothing will be foreign to us'—so that if you do not aid us, we at least may find room to print and meet your future criticisms."[18]

16. Pearson, "Weldon," p. 35.
17. Pearson, *Galton,* 3A:241.
18. Pearson to Bateson, 19 February 1901, BPB 10.

Pearson decided to wait and publish his answer to Bateson's criticism of homotyposis in *Biometrika*. In the interval Bateson tried to win Pearson over to Mendelism. He knew Pearson would be a powerful ally. On 12 October 1901, he sent a translation of Mendel's paper to Pearson, who in reply expressed skepticism about the general applicability of Mendelian inheritance. In January 1902, Weldon published a criticism of Mendelian inheritance in *Biometrika*. Bateson and Pearson exchanged heated letters concerning it. Bateson now made a last attempt at reconciliation with Pearson. He truly wanted Pearson to be on the side of Mendelism:

I respect you as an honest man and perhaps the ablest and hardest worker I have met, and I am determined not to take up a quarrel with you if I can help it. . . .
There has probably been no discovery made in theoretical biology that we can remember which approaches Mendel's in magnitude, and the consequences it leads to. This is not a matter of opinion but certain. You have worked well in the same field and if through any fault of mine you were to be permanently alienated from the work that is coming, I should always regret it. With Weldon it is different. He is a naturalist. He goes in with his eyes open. Besides, as between him and me it is too late. It was a bitter grief to me when he first made it clear to me that all partnership between us was at an end. At different times, as perhaps you know, we have each tried to renew our intercourse if not friendship, but it came to nothing and it is no use trying again. There are faults of temperament on both sides. In this matter he is now committed. How far he has mistaken not only Mendel's work but the gravity of the issue cannot be long unknown.[19]

Pearson replied:

I think sometimes you cannot be aware that Weldon has been for many years past one of my closest and most valued friends; that I do not readily make friends, and that when I say a man is my friend I mean that I have tested the strength of his affection in the graver matters of life, and am prepared to do for him and to accept from him anything that one human being can or will do for another. I think, as I say, that you have not known this, or possibly your references to him,

19. Bateson to Pearson, 13 February 1902, ibid.

—only three or four, but my memory is very jealous in such matters—would have been more guarded. As to the scientific side of the present controversies, I am perfectly ready to hear both sides, and will willingly reserve space in Part III of *Biometrika* for your defence of Mendel, if you think our Journal a suitable *locus* for your paper." [20]

This exchange between Bateson and Pearson illuminates the whole conflict between the Mendelians and biometricians. For it is evident that personality clashes were as important as scientific arguments in sustaining the conflict. If Weldon had adopted Mendelian inheritance, instead of opposing it, Pearson's whole attitude toward Mendelism might have been different. If Pearson, Weldon, and Bateson had worked together, population genetics might have begun in earnest fifteen years sooner than it did.

The breach between Bateson and Pearson soon became wider. In April of 1902 Pearson finally published a long reply to Bateson's criticisms of homotyposis. He attacked Bateson's loose definitions ("My own strong opinion is that biological conceptions can be accurately defined"), his lack of mathematical understanding, and especially his theory of discontinuous evolution ("Let me state once and for all that I consider Mr. Bateson's peculiar theory of evolution by discontinuous variations untenable").[21] Bateson's reply was equally caustic. He and Pearson were permanently at odds.

The homotyposis controversy did not directly involve Mendelian heredity. But it did raise powerful emotions which helped polarize Mendelism and discontinuous evolution on one side, from biometry and Darwinian evolution on the other.

THE MUTATION THEORY

Hugo de Vries (1848–1935) turned his attention to problems of heredity in the late 1880s. His book *Intracellular Pangenesis* was published in 1889. De Vries, believing his theory of heredity was derived from Darwin's, used Darwin's term

20. Pearson to Bateson, 15 February 1902, ibid.
21. Karl Pearson, "On the Fundamental Conceptions of Biology," *Biometrika* 1:324, 331.

for the process of inheritance. Actually de Vries's conclusions challenged the foundation of Darwinian pangenesis. Darwin conceived his theory of pangenesis primarily to account for the production of heritable individual differences, the raw material upon which selection acted, and specifically allowed for the inheritance of environmentally acquired characters.

De Vries saw Darwin's theory of pangenesis as being composed of two propositions:

1. In every germ-cell . . . the individual hereditary qualities of the whole organism are represented by definite material particles. These multiply by division and are transmitted during cell-division from the mother cells to the daughter cells.
2. In addition, all the cells of the body, at different stages of development, throw off such particles; these flow into the germ-cells, and transmit to them the qualities of the organism, which they are possibly lacking.[22]

The first of these propositions was the basis for de Vries's theory of heredity. The second, de Vries rejected because he did not think environmentally induced variations were inherited, as Weismann had proved with mutilations in mice. But by rejecting the second part of Darwin's hypothesis, de Vries eliminated the major mechanism for the production of individual differences, the raw material for selection. It is therefore hardly surprising that de Vries's revision of Darwin's idea of pangenesis led directly to his revision of Darwin's idea of evolution.

De Vries's theory of pangenesis contained two major propositions: (1) The hereditary characters of a species were mutually independent. If, said de Vries, "the specific characters are regarded in the light of the theory of descent it soon becomes evident that they are composed of single factors more or less independent of each other." The independence of specific characters was "verified in a striking manner by experiments in hybridization and crossing." [23] (2) For each hereditary character of a species there existed in the germ cell

22. Hugo de Vries, *Intracellular Pangenesis,* trans. C. Stuart Gager (Chicago: Open Court, 1910), p. 5.
23. Ibid., pp. 11, 27.

a definite material particle which determined that character. These material particles de Vries named "pangens."

It followed from these propositions that variability was of two kinds. First, the pangens might vary in their relative number: they might become more or less numerous, or change into different combinations by hybridization. Second, a pangen might, in the process of division, give rise to an altered pangen which could become active when sufficiently numerous. The first kind of variation explained Darwin's individual differences. The second explained "new characters," such as those which appeared in sports.

In 1892, three years after the publication of *Intracellular Pangenesis,* de Vries began to hybridize plants in order to trace the independent characters in subsequent generations. Between 1894 and 1899 he became convinced that the evolution of species depended primarily upon the variations caused by alteration in the pangens. The other kind of variation, which caused only individual differences, he believed unimportant for species change. By 1899 he had observed many examples of "mutations" in his stocks of *Oenothera Lamarckiana.*

In July of 1899 de Vries traveled to England for the Horticultural Society's International Conference on Hybridization, where he met Bateson. They immediately became friends, not only because they were both interested in experimental hybridization but also because they both advocated discontinuous evolution. In addition both disliked the biometricians. Bateson wrote his wife from the meeting that "de Vries is a really nice person. . . . He is an enthusiastic discontinuitarian and holds the new mathematical school in contempt—so we hit it off in admiration."[24] Bateson was delighted with the international acclaim de Vries received with his rediscovery of Mendelism in 1900.

De Vries had been working on his theory of discontinuous evolution for several years and in 1900 he finished the first volume of his *Mutationstheorie.* He denied that selection

24. BPB 1.

alone was effective for the creation of new species and propounded his theory of evolution by mutation, giving examples from his stocks of *Oenothera*. On 18 October 1900, de
Vries sent an advance copy to Bateson along with a letter
which said: "I have now the pleasure of offering you my
work on the origin of species, as discontinuous as you could
hope it." [25] De Vries fully expected an outcry from Darwinists everywhere, especially from the biometricians in England.
He wanted Bateson to join with him to present a solid front.
As de Vries stated it in a letter to Bateson, "there must be no
discontinuity between us, not even in the use of the word." [26]

The biometricians were indeed annoyed by this new attack
upon Darwinian evolution. Weldon prepared a critical paper
which was published in *Biometrika* in April 1902. He challenged de Vries's experimental proof that selection was incapable of changing a species, thus opening the door for
Darwinian evolution. As for the positive examples from the
Oenotheras, to which de Vries devoted one half of his large
volume, Weldon used only one sentence for refutation. De
Vries claimed that the offspring of most of his *Oenothera*
mutants regressed to a new center of regression, but Weldon
said he could not "find evidence that in any one of these
numerous experiments the kind of regression ascribed to the
offspring of mutations has actually occurred." [27] Weldon's
statement was nonsensical since de Vries stated that seven of
his mutant *Oenotheras* bred "absolutely constant," meaning
the offspring necessarily regressed to the new type. Weldon
concluded his argument with the statement that when regression "is better understood than it is at present such naturalists as Professor de Vries and Mr. Bateson will abandon
their attempts to distinguish between 'variations' and 'mutations.' " [28]

Weldon did not bother to send de Vries a copy of his
criticism. When de Vries did read the criticism, he was sur-

25. BPB 15.
26. De Vries to Bateson, 25 October 1900, ibid.
27. W. F. R. Weldon, "Professor de Vries on the Origin of Species,"
Biometrika 1:373.
28. Ibid., p. 374.

prised Weldon attacked the evidence from the *Oenotheras* so feebly. Writing to Bateson about Weldon's criticism, de Vries said:

> Weldon names at the end such biologists as Bateson, and de Vries, and I was glad, when reading this, to take leave from him in such good company. If you will defend me against him I will be much indebted to you.[29]

Bateson had duels with Weldon after this, but he was not motivated by the defense of de Vries. He did help de Vries become a foreign member of the Royal Society and even offered to supervise an English translation of the *Mutationstheorie*. But the relationship between de Vries and Bateson cooled. De Vries had rediscovered Mendel's work; yet Bateson had become the champion of Mendelism while de Vries was finding Mendelian inheritance of little importance in the evolution of species. On 30 October 1901, de Vries implored Bateson:

> I prayed you last time, please don't stop at Mendel. I am now writing the second part of my book which treats of crossing, and it becomes more and more clear to me that Mendelism is an exception to the general rule of crossing. It is in no way *the* rule! It seems to hold good only in derivative cases, such as real variety-characters.[30]

Bateson became more impressed with the importance of Mendelian inheritance as de Vries became less so. And as de Vries became disenchanted with Mendel, Bateson became disenchanted with de Vries.

The impact of de Vries's *Mutationstheorie* upon biologists was enormous. For many reasons biologists had become disillusioned with Darwin's idea of natural selection, and de Vries presented the first experimental evidence to support another view of the mechanism of evolution. Many biologists accepted de Vries's new theory outright, and the response was generally favorable. There were, to be sure, many old-guard Darwinists who retained their ideas. But the idea of evolution

29. De Vries to Bateson, 12 May 1902, BPB 36.
30. BPB 15.

in the first decade of the twentieth century was dominated by the surge of interest in the mutational leaps of de Vries.

The effect of de Vries's mutation theory was heightened by the growing interest in Mendelian heredity, which was demonstrated so many times with discontinuous characters between 1900 and 1910. The connection between Mendel's discontinuous variations and discontinuous evolution, although not emphasized by de Vries himself, was made by many other biologists. Many of the important adherents of Mendelian heredity during these years were also adherents of discontinuous evolution.

Many scientists thought Mendelism was necessarily associated with discontinuous evolution and was therefore anti-Darwinian. Pearson and Weldon believed this, and believed it indicated that Mendelian heredity was lacking. Pearson stated:

> To those who accept the biometric standpoint, that in the main evolution has not taken place by leaps, but by continuous selection of the favourable variation from the distribution of the offspring round the ancestrally fixed type, each selection modifying *pro rata* that type, there must be a manifest want in Mendelian theories of inheritance. Reproduction from this standpoint can only shake the kaleidoscope of existing alternatives; it can bring nothing new into the field. To complete a Mendelian theory *we must* apparently associate it for the purposes of evolution with some hypothesis of "mutations." The chief upholder of such an hypothesis has been de Vries. . . .[31]

Because Pearson and Weldon thought Mendelism was necessarily associated with discontinuous evolution, they opposed Mendelism vigorously.

Curiously, Bateson, in an argument for discontinuous evolution in 1904, stated that "when the unit of segregation is small, something mistakeably like continuous Evolution must surely exist."[32] The history of population genetics might

31. Pearson, "Weldon," p. 39.
32. William Bateson, "Presidential Address to the Zoological Section, British Association. Cambridge Meeting, 1904," in B. Bateson, *William Bateson,* p. 238.

have been accelerated had Bateson, Pearson, and Weldon taken this remark to heart.

INHERITANCE IN PEAS

Weldon initiated the attack upon Mendelian inheritance in the second number of *Biometrika* (January 1902) with an article entitled "Mendel's Laws of Alternative Inheritance in Peas."[33] He first divided inheritance into three kinds: blended, particulate (or mosaic), and alternative. Mendelian inheritance, according to Weldon, pertained only to alternative inheritance. This was not, however, the intention of Mendel, who said that his theory could account for an almost continuous array of variations. After this initial misrepresentation of Mendel's ideas, Weldon went on to attack Mendel's "law of dominance," his "law of segregation," and his neglect of ancestry.

The first general result of Mendel's work, stated Weldon, was the law of dominance. He produced examples which indicated this law was not universally true in peas, even for the characters used by Mendel, and was therefore useless. Mendel's second result was what Weldon termed the "law of segregation." This law, he claimed, was true only in very specialized cases—an accusation which was in accordance with the evidence then available to Weldon because so few experiments had been conducted and published. He concluded with the statement:

> The fundamental mistake which vitiates all work based upon Mendel's method is the neglect of ancestry, and the attempt to regard the whole effect upon offspring, produced by a particular parent, as due to the existence in the parent of particular structural characters.[34]

Weldon did not intend for this article to start a violent controversy but it did. Bateson was incensed when he read it. He had just submitted the first report concerning Mendelian heredity to the Evolution Committee of the Royal Society and was particularly enthusiastic about Mendelism when

33. *Biometrika* 1:228–54.
34. Ibid., p. 252.

Weldon's article appeared. He immediately began to prepare a detailed refutation which was published in April 1902 with translations of Mendel's papers and his own exposition of the principles of Mendelian heredity.[35]

The little book crackled with fiery comments. Bateson said it was "with a regret approaching to indignation that I read Professor Weldon's criticism." [36] He was afraid new students of heredity might discount Mendel's ideas because of Weldon's article. He also loosed a blast at the biometrical approach:

We have been told of late, more than once, that Biology must become an *exact* science. The same is my own fervent hope. But exactness is not always attainable by numerical precision: there have been students of Nature, untrained in statistical nicety, whose instinct for truth yet saved them from perverse inference, from slovenly argument, and from misuse of authorities, reiterated and grotesque.[37]

Bateson actually prepared two even stronger statements for his preface, but Cambridge University Press suggested they be dropped. They were, but Bateson later said he "rather liked these two bits!" [38]

After close study of Weldon's arguments, it was evident to him that "Professor Weldon's criticism is baseless and for the most part irrelevant, and I am strong in the conviction that the cause which will sustain damage from this debate is not that of Mendel." [39] He proceeded to refute Weldon's assertion that Mendelian inheritance was applicable only to alternative inheritance. He then challenged Weldon's belief that Mendel had propounded a law of dominance and made a careful attack upon every shred of evidence Weldon had utilized.

In defense of Mendel's law of segregation, Bateson expounded the "purity of the germ cells." He cited experiments which showed that extracted recessives were "identical" to

35. William Bateson, *Mendel's Principles of Heredity: A Defence* (Cambridge: Cambridge University Press, 1902).
36. Ibid., p. vi.
37. Ibid., p. x.
38. William Bateson to Beatrice Bateson, 28 April 1902, BPB 26.
39. Bateson, *Defence,* p. 108.

their recessive grandparents, a phenomenon which could not be explained by Pearson's law of ancestral heredity. Bateson's choice of words here was unfortunate because segregation of other factors might make the organism with the extracted recessive distinctly different from either grandparent. Weldon later insisted that Bateson believed any organism with an extracted recessive factor must be exactly like one of its ancestors. Bateson's exposition of the "purity of the germ cells" also ignored interaction effects, which he discovered only later.

The tone of Bateson's *Defence* made the biometricians unhappy. In his memoir of Weldon, Pearson stated that "Mr. Bateson's defence deeply pained Weldon, and rendered it difficult for a finely strung temperament to maintain—as it did to the end—the impersonal tone of scientific controversy." [40] The truth was that Weldon was scarcely less "impersonal" than Bateson, and Pearson's own personal attacks on Bateson and others were virulent.

Weldon could scarcely reply to Bateson's able defense of Mendel's ideas. Bateson, after all, was at this time perhaps the foremost expert in the world on Mendelian heredity. Instead of a reply to Bateson's criticisms, Weldon started a new attack. In his next paper, "On the Ambiguity of Mendel's Categories," published in November 1902, Weldon first stated that he could "see no reason to modify the statements" [41] he had earlier made about Mendelian heredity. He went on to challenge the accuracy with which a Mendelian character might be classified. Mendel had said only that when green peas are crossed with yellow the hybrid seeds were green— the shade had not been specified. Weldon gave examples, some drawn from Bateson's work, where supposed Mendelian categories were inexact and concluded that the ancestral law of inheritance might be operating.

The Mendelians reacted to Weldon's criticism by making certain the characters they used in breeding experiments were distinct. Bateson insisted upon this. The effect of Weldon's criticism was to delay the analysis of continuously varying

40. Pearson, "Weldon," p. 42.
41. *Biometrika* 2:44.

characters in terms of Mendelian inheritance because experimenters wanted clear-cut characters.

The controversy over inheritance in peas ended, not because the antagonists were satisfied but because new controversies concerning Mendelian heredity had come to the fore.

HEREDITY IN MICE

In the first report to the Evolution Committee, Bateson suggested that Mendelian ratios were to be found in mice.[42] Before reading this report Weldon claimed in his article "Mendel's Laws of Alternative Inheritance in Peas" that inheritance in mice did not follow Mendel's laws and must be explained by the law of ancestral heredity. Bateson of course replied to this charge in his *Defence*. Weldon had unfortunately utilized the work of the German biologist Johann von Fischer, whom he quoted as an "excellent authority." Bateson showed that von Fischer's claims were outrageous and that Weldon was making no distinction between "wild type" hybrids and "wild type" pure breds.

Weldon decided to begin breeding experiments with mice which would prove beyond a doubt that Mendelian heredity could not account for the results. He encouraged his pupil, A. D. Darbishire, to proceed with the breeding experiments. Darbishire crossed the Japanese waltzing mouse with the common albino mouse. In his first of four reports,[43] published in November 1902, Darbishire recorded results from the first nine crosses. The hybrids showed four different coat patterns, so Darbishire concluded that Mendel's law of dominance did not hold. Furthermore, the offspring of inbred albinos showed less white than the offspring of albinos which had appeared in litters of piebald mice. Darbishire concluded that although "on the Mendelian hypothesis the ancestry of the albinos should make no difference: we shall see that, as a matter of

42. *Reports to the Evolution Committee of the Royal Society,* Report 1, Experiments Undertaken by W. Bateson, F. R. S., and Miss E. R. Saunders (London: Harrison and Sons, 1902), p. 145.
43. A. D. Darbishire, "Note on the Results of Crossing Japanese Waltzing Mice with European Albino Races," *Biometrika* 2:101–4.

fact, it probably does." [44] Weldon continually used the argument that according to Mendel's hypothesis any factor would be expressed the same way no matter what the other factors in a gamete were, and Darbishire adopted this argument.

Bateson was intensely interested in the experiment and corresponded often with Darbishire. In a letter dated 3 January 1903, Darbishire wrote Bateson, "I am absolutely unbiased about Mendel and am very keen to come to an unprejudiced conclusion on it." [45] But Darbishire's reports showed, and he later admitted, that he was definitely prejudiced against Mendelian heredity at this time.

When the first report was published, Bateson was immediately suspicious. He wrote Darbishire, whom he found had neglected to mention that none of the hybrids were waltzers and that all had dark eyes, even though both parents had pink eyes. This, Bateson wrote Darbishire, looked like Mendelian inheritance.

Darbishire's second report appeared in February 1903. It described the hybrids from twenty pairings and the initial results of pairing hybrids with hybrids and hybrids with albinos. The first-generation hybrids were not uniformly colored, indicating to Darbishire that coat color was not subject to Mendelian inheritance: "any modification of Mendel's hypothesis involves the uniformity of the first generation." Darbishire admitted that the appearance of albinos and waltzers was so far "in possible accordance with some form of Mendelian hypothesis." He stated flatly that "the inheritance of eye color is not in accordance with Mendel's results," because pure pink-eyed parents had produced dark-eyed young. [46]

Bateson in response wrote a letter to *Nature* in which he suggested that the initial strains might not really be pure, although they bred true for the waltzing and albino characters, which would account for the variation of coat color in the hybrids. He also proposed a simple Mendelian interpretation

44. Ibid., p. 102.
45. BPB 27.
46. A. D. Darbishire, "Second Report on the Result of Crossing Japanese Waltzing Mice with European Albino Races," *Biometrika* 2:170, 172, 174.

of the eye color results. Weldon reacted with a letter challenging Bateson's interpretations. A series of letters between Bateson and Weldon followed, until the editor of *Nature* refused to publish anything more on the subject.

At this time Darbishire prepared his third report; Weldon, his next attack upon Bateson's Mendelism; and Bateson, an article on color heredity in rats and mice. Darbishire's third report appeared in June 1903 and began with the familiar biometrical refrain:

It is an essential part of the Mendelian hypothesis that the (so-called "extracted") recessive individual which is produced by pairing two first crosses, is in every respect similar to the original pure recessive. It forms, in fact, the foundation on which the doctrine of the purity of the germ cells rests.[47]

Bateson had by now denied this many times. Darbishire, surely with Weldon's encouragement, used the argument once more to show that the results with the mice were not Mendelian.

Weldon's paper, entitled "Mr. Bateson's Revisions of Mendel's Theory of Heredity,"[48] appeared in the same number of *Biometrika*. He argued that when a situation which did not fit Mendel's theory arose, Bateson simply revised Mendel's ideas until they fit. This meant Mendel's ideas could explain anything, and therefore nothing. In the light of later developments in genetics, Weldon's arguments were most unfortunate. He attacked Bateson's beliefs that: (1) dominance was unessential for Mendelian inheritance; (2) atavism could be explained by Mendelian inheritance; (3) sex linkage of characters exists; and (4) sex was a Mendelian factor. In the last part of the paper Weldon attacked Bateson's revisions of Mendel as applied to Darbishire's results and decided that they were an ineffective explanation of the data.

Bateson's article "The Present State of Knowledge of Colour-Heredity in Rats and Mice"[49] was written independently

47. A. D. Darbishire, "Third Report on Hybrids between Waltzing Mice and Albino Races," *Biometrika* 2:282.

48. *Biometrika* 2:286–98.

49. *Proceedings of the Zoological Society of London* 2 (1903); reprinted in Bateson, *Scientific Papers*, 2:76–108.

of Weldon's. He proposed Mendelian methods for the analysis of coat color in mammals. The article answered the major points raised by Weldon and Darbishire; so Bateson rested his case until Darbishire's final results were published.

Darbishire published his major results and conclusions in January 1904. He admitted now that albinism segregated in Mendelian ratios, as did eye color, and that waltzing was completely recessive. But he went on to present what he considered to be grave challenges to further Mendelian interpretation. First, albinism was not a true recessive because variable offspring appeared in the first-generation hybrids. Second, although behaving as a recessive, waltzing did not segregate in Mendelian ratios. Darbishire's argument on this point is worth quoting because he was to retract the conclusion less than three months later:

> *Waltzing* occurs in only 97 out of the 555 individuals resulting from the union of hybrids. When we compare this with the number of pink-eyed individuals (131–134) or of albinos (137) we see that the proportion of waltzing individuals cannot be regarded as a possible quarter. . . . the odds against so great a deviation being rather more than 50,000 to 1. . . . The evidence that the waltzing character does not segregate in Mendelian proportions is very strong.[50]

Darbishire produced other data which contradicted the Mendelian interpretation—data he called "the most conclusive results which I have obtained." [51] He bred together hybrids which he claimed were gametically the same but with differing amounts of albino ancestry. If purebred waltzers are denoted by W, purebred albinos by A, and hybrids by H, the three crossings Darbishire made are given in figure 1. The Mendelian interpretation would be that each of these crosses would produce equal numbers of albinos, whereas Galton's law of ancestral heredity would predict that the crosses with mice of greater albino ancestry would produce greater numbers of albinos. Darbishire's data strikingly confirmed the

50. A. D. Darbishire, "On the Result of Crossing Japanese Waltzing with Albino Mice," *Biometrika* 3:20.
51. Ibid., p. 23.

law of ancestral heredity. The gist of his entire paper was that the Mendelian interpretation of heredity in mice counted for little and that the law of ancestral heredity counted for much.

Bateson immediately began to correspond with Darbishire, asking critical questions about the way the data was derived and interpreted. He pointed out numerous inconsistencies between Darbishire's fourth paper and his other papers, and

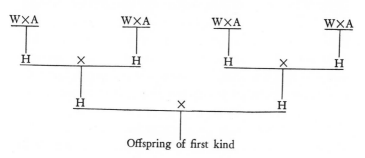

Offspring of first kind

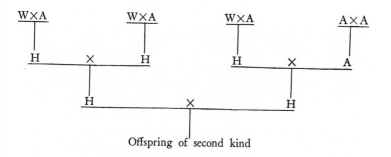

Offspring of second kind

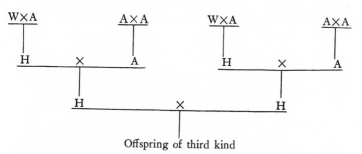

Offspring of third kind

Fig. 1. Darbishire's crosses of hybrids with differing amounts of albino ancestry.

within the fourth paper itself. Soon he convinced Darbishire of two important points: that waltzers were less viable and therefore did appear in Mendelian ratios in his results, and that new pure-breeding varieties could arise discontinuously by means of Mendelian recombination. In a paper delivered 15 March 1904, Darbishire said that in the hybrid offspring (F_2) of his experiment the Mendelian expectation was 25 percent waltzing mice: "this is very roughly what happens." [52] Later in the same paragraph he said "one in every four is a waltzer." To have reversed his rejection of Mendelian segregation with the waltzers was insult enough to the biometricians, but he also claimed to have produced albino waltzers. Since albinism bred pure and waltzing did too, Darbishire concluded that he had produced a new pure-breeding variety in a discontinuous leap.

Weldon and Pearson were irritated and made Darbishire aware of their unhappiness. The situation was difficult for Darbishire because he depended upon Weldon for recommendations to teaching positions. To make matters worse, Bateson now made the startling discovery that Darbishire, in his crosses of hybrids which proved that the law of ancestral heredity accounted for the phenomena better than the Mendelian hypothesis did, had not distinguished between pure-bred dominants and hybrids. The diagram of matings (fig. 1) shows why Darbishire's failure to make this distinction would lead to fewer albinos in the first and second crosses than predicted by the Mendelian hypothesis. Darbishire's most conclusive results were blatantly invalid. Moreover, Bateson's investigations now cast grave doubts about the accuracy of Darbishire's records. He revealed this to Darbishire in a letter dated 22 May 1904.[53] Darbishire, shaken by Bateson's discoveries, was in a very awkward position. He had already incensed Pearson and Weldon, and now Bateson was about to reveal the depths of his mistakes in his best pub-

52. A. D. Darbishire, "On the Bearing of Mendelian Principles of Heredity on Current Theories of the Origin of Species," *Manchester Memoirs* 48, no. 24 (1904): 13.
53. BPB 27.

lished scientific research. His reputation as an investigator was at stake.

He wrote Bateson a desperate letter in which he tried to arrange a secret meeting with him to put the records in order. He pleaded with Bateson not to make his discoveries public and asked for help:

> I hope you will do your best to get me out of the position I am in as soon as possible and I pray you not to mention this letter to anyone. What do you suggest?
> I don't mind your saying what you like about the interpretations and conclusions in the mouse paper; but to have my records discredited would be heart-breaking and render it useless and a waste of time for me to go on with the costly experiments I am carrying out now.[54]

Bateson replied: "It will, I think, be obvious to you on reflexion, that any communication between us which is to serve as a basis of discussion must be of a public nature."[55] But Bateson made no public disclosures about the inadequacies of Darbishire's work because William Castle in the United States was doing that job and because Darbishire was being scared into the hands of the Mendelians. Bateson did not wish to ruin a good thing.

At the meeting of the British Association on 18 August 1904, Darbishire stated his opinion on behalf of the Mendelians that waltzing was a recessive which segregated roughly in accordance with Mendelian expectation. Pearson was galled at this public change of view by Darbishire. On 29 September 1904, he wrote a scathing letter to *Nature* in which he reproduced Darbishire's earlier and later views side by side, with the comment:

> Which writer shall a member of the inquiring general public trust? Or, if the two writers should be the same, must we assume that in Oxford, under the influence of some recessive biometer [Weldon], Mr. Darbishire failed to see that 97 in 555 was a reasonable quarter, or 20 in 555 a reasonable sixteenth, but that he has learnt in Manchester, or perhaps in

54. Darbishire to Bateson, 27 May 1904, ibid.
55. Bateson to Darbishire, 30 May 1904, ibid.

Cambridge from some dominant anaesthetist [Bateson], that these things really are so? [56]

Having been subjected for two years to Bateson's criticism, and now to that of the biometricians, Darbishire sought to mediate. In a paper delivered on 10 January 1905, he attempted to show that the Mendelian and biometric approaches were not contradictory.[57] He retracted his claim that the results of his mice breeding led necessarily to the law of ancestral heredity with the explanation that he had not chosen his hybrids properly. The biometric and Mendelian interpretations were not contradictory, he said, because each had a particular point of view. They were like skew lines which did not intersect. This paper represented a transition stage for Darbishire, who became an avowed Mendelian. In 1911 he published an influential text entitled *Breeding and the Mendelian Discovery*. In it he lamented his training in heredity as received from Weldon.

MENDELISM AND BIOMETRY

In his first paper on Mendel, Weldon claimed that Mendelian heredity ignored "ancestry," so the laws of Mendel and the laws of ancestral heredity of Galton and Pearson were incompatible. In his *Defence*, written in response to Weldon's paper, Bateson agreed emphatically that Mendelian heredity and the law of ancestral heredity were incompatible. He declared that "the Mendelian principle of heredity asserts a proposition absolutely at variance with all the laws of ancestral heredity, however formulated." [58] Weldon and Bateson attempted to show the incompatibility of the two theories of heredity because they believed that Mendelism was associated with discontinuous evolution and the laws of ancestral heredity with Darwinian evolution. This belief, a consequence of the arguments which preceded the rediscovery of Mendelism,

56. *Nature* 70 (1904): 530.
57. A. D. Darbishire, "On the Supposed Antagonism of Mendelian to Biometric Theories of Heredity," *Manchester Memoirs* 49, no. 6 (1905): 1–19.
58. Bateson, *Defence*, p. 114.

was detrimental to the synthesis of Mendelism and Darwinism.

A few were unconvinced that Mendelism and the law of ancestral heredity were incompatible. In response to Bateson's *Defence*, the British mathematician G. Udny Yule wrote a long article debunking Bateson's reasons for asserting this incompatibility. After criticizing Bateson's militant tone in the *Defence*, Yule said that the law of ancestral heredity had been applied to intraracial heredity whereas Mendelism so far had been applied to hybridization only; therefore, they could not be said to be contradictory.

Yule considered Mendel's hypothesis for explaining his results to be "ingenious and remarkable." He showed that, assuming complete dominance, the randomly bred offspring of the hybrid generation would maintain the 3:1 proportion indefinitely.[59] In the case of the dominant characteristic, the results predicted by Mendelian heredity were precisely the same as those predicted by the law of ancestral heredity, in the general sense that the chance of an organism with the dominant characteristic A producing offspring with A is increased if the ancestry also exhibits A. Yule concluded that "Mendel's Laws, so far from being in any way inconsistent with the Law of Ancestral Heredity, lead directly to a special case of that law." [60]

Yule knew that for a recessive trait Mendel's predictions and the law of ancestral heredity did not agree. So he asked: "In what way may the special conditions under which Mendel's Laws hold good be broadened so as to permit of a generalization of the results?" [61] Yule suggested that the assumption of complete dominance should be dropped and the effect of the environment upon the expression of gametic characters should be taken into account. If these two modifications were assumed, he showed mathematically that the predictions of

59. This was a special case of the Hardy-Weinberg law. Yule at this time believed the 3:1 ratio was the only stable equilibrium.
60. G. Udny Yule, "Mendel's Laws and Their Probable Relations to Intra-Racial Heredity," *New Phytologist* 1:226–27.
61. Ibid., p. 227.

Mendel's theory and the law of ancestral heredity could be consistent.

Yule assailed the belief (held by Bateson, Weldon, and Pearson) that Mendelism was necessarily associated with discontinuous evolution. He suggested the multiple factor hypothesis of apparently continuous variation and the possibility that Mendelian factors might themselves be variable in small but discontinuous steps, as in the slight change of a large molecule. Since Mendelism could account for continuous variations, it was compatible with biometry and Darwinian evolution. Yule concluded with the thought that it was

essential, if progress is to be made, that biologists—statistical or otherwise—should recognise that Mendel's Laws and the Law of Ancestral Heredity are not necessarily contradictory statements, one or other of which must be mythical in character, but are perfectly consistent the one with the other and may quite well form parts of one homogeneous theory of heredity.[62]

Yule's excellent paper had little effect upon the widening gap between the Mendelians and biometricians. Not until R. A. Fisher's first genetical paper in 1918 was there an important attempt in England to follow the lead suggested by Yule.

Now it was Karl Pearson's turn. In 1904 he published a paper entitled "On a Generalized Theory of Alternative Inheritance, with Special Reference to Mendel's Laws." [63] He explored the mathematical consequences of the pure gamete theory, namely, that characters are inherited intact. He checked to see if the predictions of the pure gamete theory as defined by Mendel were in accordance with observations already made by the biometricians.

First Pearson found that (what is now called) the Hardy-Weinberg equilibrium was a necessary consequence of the pure gamete theory:

However many couplets we suppose the character under investigation to depend upon, the offspring of the hybrids—or

62. Ibid., p. 236.
63. *Philosophical Transactions of the Royal Society,* A, 203 (1904): 53–86.

the segregating generation—if they breed at random *inter se,* will not segregate further, but continue to reproduce themselves in the same proportions as a stable population.[64]

He went on to make the extraordinary argument that sexual reproduction on the Mendelian scheme produces little novel heritable variation:

It is thus clear that the apparent want of stability in a Mendelian population, the continued segregation and ultimate disappearance of the heterozygotes, is solely a result of self-fertilization; with random cross fertilization there is no disappearance of any class whatever in the offspring of the hybrids, but each class continues to be reproduced in the same proportions. Thus our generalized theory lends no countenance to the appearance of any "mutations" within a hybrid population under random mating; the only appearance of new constitutions is in the segregating generation, or the first generation of hybrid offspring. Except at this stage, the appearance of the unfamiliar is only the chance occurrence of a very rare normal variation. When we recollect that a purely allogenic [homozygous] individual is only to be expected once in a population of 4^m individuals, or if there be ten couplets, once in more than a million individuals, it will be clearly seen that the variety of some of the more exceptional normal constitutions may easily lead to their being looked upon as "mutations," even if they appear in the offspring of a population many generations removed from hybridization.[65]

Pearson, without realizing it, had pointed out a huge source of heritable variation as a consequence of genetic recombination. He could only see that Mendelian heredity produced no "mutations." H. Nilsson-Ehle, Edward East, and other geneticists later used Pearson's same reasoning to argue that Mendelian heredity provided most of the heritable variability in a population. Pearson was so near and yet so far from being able to harmonize his idea of Darwinian selection with his idea of Mendelian heredity. Wilhelm Johannsen used the same reasoning to argue that species change must occur by large mutational leaps, as de Vries claimed (see chap. 4).

Pearson found that the mathematical consequences of a

64. Ibid., p. 60.
65. Ibid.

pure gamete theory were in accordance with his researches into heredity, except that the theoretical values on Mendelian assumptions, including complete dominance, for the phenotypic correlation between parent and offspring ($\frac{1}{3}$), between brothers (0.3 to 0.4), and between grandparent and offspring ($\frac{1}{6}$) were well below the observed values of 0.5, 0.5, and 0.2 to 0.3. Although no inherent inconsistency existed between Mendelism and the biometric description of inheritance in populations, the pure gamete theory, said Pearson, was "not elastic enough to account for the numerical values of the constants of heredity hitherto observed." [66]

Pearson thus rejected Mendelian heredity for the time. He did say that assortative mating or incomplete dominance would change the correlations he had derived with the pure gamete theory, but evidence was lacking about these processes. He suggested that the Mendelians produce "a few simple general principles . . . which embrace *all* the facts deducible from the hybridization experiments of the Mendelians; these can form the basis of a new mathematical investigation." [67] Unfortunately, it was obvious that Pearson himself would not care to undertake such an investigation.

Bateson had no way to reply to Pearson's statistical criticisms of Mendelism, but two years later Yule rebutted Pearson's calculations.[68] Pearson had assumed, as a "generalization" of Mendelian heredity, that only one type of homozygote determined the character upon which the correlations were calculated—which enabled him to avoid the whole question of dominance. But his claim to having avoided assumptions regarding dominance was misleading. Yule showed that Pearson's method was mathematically equivalent to the assumption of complete dominance in the correlation equations. He went on to say, as he had in 1902, that if incomplete dominance and environmental effects were taken

66. Ibid., p. 86.
67. Ibid.
68. G. Udny Yule, "On the Theory of Inheritance of Quantitative Compound Characters on the Basis of Mendel's Laws—A Preliminary Note," in *Report of the Third International Conference on Genetics* (London: Spottiswoode, 1907), pp. 140–42.

into account, the Mendelian interpretation could account for the correlations measured in populations by Pearson and his colleagues.

Yule was ahead of his time. In 1906 he was probably the only biometrician in England who recognized not only that Mendelism and biometry were compatible but also, even more crucial, that Mendelism and Darwin's idea of continuous evolution were compatible. It is true that in 1905 and 1906, Darbishire, who had been battered by both the Mendelians and biometricians, published papers which said no conflict between Mendelism and biometry existed. But Darbishire believed they did not conflict because their points of view were so different; he did not advocate the synthesis of Mendelism and biometry, as Yule had done.

MEETING OF THE BRITISH ASSOCIATION, 1904

Perhaps the most heated and publicized debate between the biometricians and Mendelians occurred at the meeting of the zoology section of the British Association, 18 and 19 August 1904. Bateson, then president of the section, planned his address and organized his fellow Mendelians with the intention of scoring a crushing public victory over the biometricians. He began work on it in June and made certain his colleagues would have their best evidence prepared. Weldon boned up on all his criticisms of Mendelism in preparation for the meeting.

On the morning of 18 August, Bateson delivered his militant challenge to the biometricians. He lauded Mendelian investigations and discontinuous evolution, challenged the Darwinian selection theory, and directly confronted the biometricians. Speaking of the careful process of domestic breeding, which often utilized discontinuous variations, Bateson stated:

Operating among such phenomena the gross statistical method is a misleading instrument; and, applied to these intricate discriminations, the imposing Correlation Table into which the biometrical Procrustes fits his arrays of unanalysed data is still no substitute for the common sieve of a trained judgment.[69]

The next morning the Mendelians began to present their data. Miss E. R. Saunders presented material on inheritance in plants; Darbishire presented his researches on mice with his newly acquired bias toward the Mendelian interpretation; C. C. Hurst spoke on heredity in rabbits. Then Weldon opened the discussion and raised four or five of his choice arguments against Mendelism. He concluded his remarks with the comment, as summarized by *Nature,* that

> until further experiments and more careful descriptions of results were available, it was better to use the purely descriptive statements of Galton and Pearson than to invoke the cumbrous and undemonstrable gametic mechanism on which Mendel's hypothesis rested.[70]

The afternoon meeting promised to be lively. Punnett, speaking in 1949, recalled the action:

> We adjourned for lunch and on resuming found the room packed as tight as it could hold. Even the window sills were requisitioned. For the word had gone round that there was going to be a fight. Probably other meetings were depleted— but after all the Association is British. Weldon spoke with voluminous and impassioned eloquence, beads of sweat dripping from his face, and I cannot help recalling the admiring remark made by one young Oxford man to another as they sat just in front of me, "Clever beggar that—he hasn't got to stop and think." Bateson replied and there may have been other speakers, I have forgotten. But towards the end Pearson got up and the gist of his remarks was to propose a truce to controversy for three years, after which the protagonists might meet again for further discussion. On Pearson resuming his seat, the Chairman, the Rev. T. R. Stebbing, a mild and benevolent looking little figure for a great carcinologist, rose to conclude the discussion. In a preamble he deplored the feelings that had been aroused, and assured us that as a man of peace such controversy was little to his taste. We all began fidgeting at what promised to become a tame conclusion to so spirited a meeting, especially when he came to deal

69. Bateson, "Presidential Address," in B. Bateson, *William Bateson,* p. 240.
70. *Nature* 70 (1904): 539.

with Pearson's suggestion of a truce. But we need not have been anxious, for the Rev. Mr. Stebbing had in him the makings of a first-rate impresario. "You have all heard," said he, "what Professor Pearson has suggested" (pause), and then with a sudden rise of voice, "But what I say is let them fight it out." And on that note the meeting ended. Bateson's generalship had won all along the line and thenceforth there was no danger of Mendelism being squelched out through apathy or ignorance.[71]

Punnett's memory was not exact: he himself spoke in the afternoon meeting, not in the morning as he recalled; also, Weldon spoke in the morning and Bateson not until afternoon. But the flavor of the victory that the Mendelians tasted that afternoon is obvious in Punnett's account, forty-five years later. As a result of this meeting Pearson attended only one more meeting of the British Association. As for Weldon, in Pearson's words, "the excitement of the meeting . . . seemed to brace Weldon to greater intellectual activity and wider plans."[72]

COAT COLOR IN HORSES

In 1900 C. C. Hurst, who had been experimenting with hybridization of orchids, became a zealous disciple of Mendelian inheritance. Hurst was, in Punnett's words, "over-apt to find the 3:1 ratio in everything he touched."[73] Because of his desire to give the biometricians no room for attack, Bateson was sometimes skeptical of Hurst's claims. In 1906, after a study of Weatherby's *General Stud Book of Race Horses,* Hurst came to the conclusion that chestnut was a simple Mendelian recessive to bay and brown. He wrote a short paper on the subject and asked Bateson to communicate the paper to the Royal Society, which Bateson did although with some reluctance.

Weldon was at this time chairman of the Zoological Committee, and Hurst's paper was submitted to him. He immediately "threw himself nine hours a day into the study of

71. Punnett, "Early Days of Genetics," *Heredity* 4 (1950): 7–8.
72. Pearson, "Weldon," p. 44.
73. Punnett, "Early Days of Genetics," p. 8.

The General Studbook," [74] where he found several examples which contradicted Hurst's thesis. These exceptions he dramatically presented after Hurst had read his paper at the meeting on 7 December 1905. Hurst, standing his ground, "blandly assured Professor Weldon that he was mistaken and that these alleged exceptions were mere errors of entry." [75] This irritated Bateson, and he withdrew Hurst's paper from publication.

Later, Hurst discovered that Weldon's most decisive cases were indeed errors of entry. He added a note to his original paper explaining this, and Bateson resubmitted the paper.[76] Weldon was outraged and continued his intensive study of the *Stud Book* in order to prove that the Mendelian interpretation did not hold. The *Stud Book* was in twenty volumes, and Weldon was still working on this material when, after a sudden illness, he died on 13 April 1906.

Pearson mourned the loss of his friend. He was angry that arguing with the Mendelians had taken so much of Weldon's time. When Hurst wrote Pearson to express his regrets about Weldon, Pearson replied:

> Only a few days before his death he [Weldon] condemned in stronger language than I have ever heard him use of any individual the tone and contents of the note added to your paper. It is a judgment in which I believe every man who has the interests of science at heart will concur.[77]

On this sour note the conflict between the Mendelians and the biometricians largely ceased. After Weldon died, Pearson redirected his interests from heredity and evolution toward the problems of practical eugenics and methods of applied statistics. He still published an occasional criticism of Mendelian interpretations, but he did not want to engage again in controversy. So in England the conflict died away.

74. Pearson, "Weldon," p. 47.
75. Punnett, "Early Days of Genetics," p. 8.
76. C. C. Hurst, "On the Inheritance of Coat Colour in Horses," *Proceedings of the Royal Society,* B, 77:388–94.
77. This quote was given by Hurst in a letter to Bateson, 9 May 1906, BPB 21.

THE EFFECTS OF THE CONFLICT

The conflict between the Mendelians and biometricians had its roots in the argument over whether evolution was continuous, as Darwin had claimed, or discontinuous, as Huxley and Galton had claimed. The net effect of the conflict was to exacerbate the argument over continuity in evolution. During the struggle Mendelism was firmly associated with discontinuous evolution and biometry with Darwinian evolution. De Vries's mutation theory had much appeal by itself; associated with Mendelian inheritance, it seemed even stronger, despite de Vries's own views. Thus initially, as Mendelism gained, Darwinism lost.

The conflict had a widespread influence. It touched biologists in the United States, Sweden, Germany, and France, as well as in England and Holland. The result was not always a split between Mendelism and biometry: the Americans C. B. Davenport and Raymond Pearl studied biometry in England and both became Mendelians, to the dismay of Pearson and Weldon. But the other effect of the conflict in England, the gulf between Mendelism and Darwinism, was wide in the United States as elsewhere. Pearl was a staunch believer in discontinuous evolution and Davenport leaned strongly in that direction.

Yule's approach of synthesizing Mendelism and biometry in the study of Darwinian evolution was submerged by the conflict. His was the approach of population genetics. Conflicts among his contemporaries prevented its development at this time.

In 1906 the mutation theory and Mendelism appeared to be on the way to a victory over Darwinian evolution. The oppositon of the biometricians had been broken and experimental evidence was fast accumulating in favor of Mendelian inheritance. But the mutation theory claimed little positive evidence and was based upon the belief that without major mutations selection was ineffective in changing a species. This belief soon faced serious challenges.

4   Darwinian Selection: The Controversy, 1900–1918

In 1906 Mendelism and Darwinism were separated by the consequences of the conflict between the Mendelians and the biometricians. Twenty-five years later Mendelian heredity and Darwinian selection were quantitatively synthesized into population genetics. Obviously, the intervening years witnessed a decline in the antagonism between Mendelism and Darwinism—a decline that may be traced directly to the results of research on the selection problem.

Darwin's study of the experience of breeders convinced him that selection was effective when acting upon small continuous variations. Alfred Russel Wallace and August Weismann, perhaps the best-known evolutionists of the late nineteenth century, believed that Darwin's conception of selection was correct. The twentieth century, however, brought vigorous attacks upon Darwinian selection. The ensuing debate over the selection problem stimulated biologists to begin selection experiments and an analysis of the heritability of continuous variations.

During the first decade after the rediscovery of Mendelism, researches into the selection problem appeared to justify the mutation theory and to undermine Darwin's natural selection theory. At this time almost all Mendelians believed in the mutation theory. But during the following decade, as a result of their own researches into the selection problem, most Mendelians came to realize that Mendelian heredity, far from being antagonistic to Darwinian selection, was in reality complementary to it. The precise reasons for this change of view were obscure even to geneticists who worked during these two decades. A. Franklin Shull, a prominent geneticist at the University of Michigan, published a popular text on evolution in 1936 which stated:

Just how, or when, natural selection began to be again more favorably regarded can not be stated. It was a gradual process, involving a change in the attitude of many biologists. To what extent its revival has already occurred it is likewise impossible to say. Such movements can not be measured.[1]

In this chapter I attempt to analyse the arguments against Darwinian selection which dominated the decade 1900–1910 and, despite Shull's pessimism, to trace the revival of selection theory in the following decade.

THE ARGUMENT AGAINST DARWINIAN SELECTION

Huxley, Galton, and Bateson all doubted whether the selection of continuous variations was as effective as Darwin imagined. Huxley and Bateson believed that continuous variations were too small to generate significant selection pressures. Galton believed that the selection of continuous variations soon reached a limit because of the counteracting effect of regression. But none of them produced sustained selection experiments to prove these points. De Vries, who gained the respect of many biologists simply because he supported his theories with concrete experimental evidence, conducted selection experiments which he believed supported Galton's predictions and which were therefore a valuable added proof for his mutation theory. De Vries thought his experiments demonstrated that selection was ineffective when operating upon continuous variations, so large mutations had to be the source of variation essential to evolution.

In the first volume of *The Mutation Theory,* in a chapter entitled "Selection Alone Does Not Lead to the Origin of New Species," de Vries presented evidence from his own experiments and those of others against the effectiveness of selection acting upon continuous variations. He concluded the following:

1. A Franklin Shull, *Evolution* (New York: McGraw-Hill, 1936), pp. 210–11.

1. Characters vary continuously only in a plus or minus direction. Thus "the character can be diminished or increased, *but nothing new can arise in this way.*" [2]

2. Selection in a population is effective for only four or five generations, and selection must be continued to maintain the population at this peak.

3. Galton was correct in thinking that regression counteracted selection. "Selection is succeeded by regression, which is great in proportion to the stringency of the selection which preceded it. However long the selection is maintained it is always followed by regression." Consequently, the characters produced by selection always revert to their original form when selection ceases; "the time it takes them to disappear is the same as it took them to appear." [3]

Weldon challenged the validity of de Vries's selection experiments. But de Vries had captured the imaginations of experimental biologists, and in 1903 new evidence published by the Danish botanist Wilhelm Johannsen gave strong support to the mutation theory.

WILHELM JOHANNSEN'S PURE LINE THEORY

Johannsen (1857–1927) set himself the task of determining the relationship of selection and Galton's law of regression in populations. Instead of choosing a species which exhibited cross-fertilization, he chose a species which reproduced by self-fertilization. The population of self-fertilizing individuals was composed of "pure lines." By a pure line Johannsen meant all the individuals which descended from a single self-fertilized individual. He believed that heredity in pure lines must be the simplest case and that if this case were understood a general theory might be based upon it. [4]

2. Hugo de Vries, *The Mutation Theory*, trans. J. B. Farmer and A. D. Darbishire, 2 vols. (Chicago: Open Court, 1910), 1:118.

3. Ibid., p. 120.

4. Johannsen's initial results in pure line work were published in Danish in 1903 by the Royal Danish Scientific Society. A German translation appeared the same year: *Ueber Erblichkeit in Populationen und in Reinen Linien* (Jena: Gustav Fischer, 1903). The most complete English translation is by Harold Gall and Elga Putschar, "Con-

Johannsen intended that his experiments distinguish between Galton's theory of selection, which had been experimentally verified by de Vries, and Darwin's theory of selection, which Pearson and Weldon promulgated. De Vries maintained that selection of continuous variations was ineffective. The biometricians believed such selection could change a population almost indefinitely.

In the spring of 1901 Johannsen bought 16,000 brown "Princess" beans, cultivated forms of *Phaseolus vulgaris*. From these he planted 100 that represented the average characters of the whole lot in regard to length, breadth, and weight. He also planted the 25 smallest and the 25 largest beans. The offspring of the largest and smallest beans deviated from the mean in the same direction but to a lesser degree than their parents, as predicted by Galton's law of regression. But Johannsen was not satisfied. He wanted to know what was happening to the pure lines during the experiment. He suspected that Galton's law of regression was invalid for a pure line.

In the spring of 1902 Johannsen commenced his pure line experiments. He chose parent beans from 19 plants of the 1901 crop. Each of these 19 plants was the outgrowth of a single carefully measured bean from the original purchased lot. Within each of the 19 pure lines, the parent (1901 crop) beans were grouped into weight classes. Later the offspring of each weight class were weighed and their average weight computed. The offspring beans (1902 crop) were also classified into weight classes with the offspring of each parent bean considered separately.

Taken together, the weights of the 1902 offspring produced a smooth random distribution with no indication that the population was composed of pure lines. Moreover, the data as a whole again confirmed Galton's law of regression. Johannsen claimed his results were even better than Galton's results with peas. This time, however, Johannsen was able to analyse

---

cerning Heredity in Populations and in Pure Lines," *Selected Readings in Biology for Natural Sciences 3* (Chicago: University of Chicago Press, 1955), pp. 172–215. The quotes are from this English translation.

the data by pure lines as well as by treating the population as a unit. The results showed *"that selection has had no reliably demonstrable influence on the types of the pure line,"* which meant that in a pure line *"the regression is complete, quite up to the type of the line.* The personal character of the mother-bean has no influence, that of the grandmother, etc., also none; but the *type of the line* determines the average character of the offspring."[5] Galtonian regression applied in the population as a whole because a selected group of extreme parent beans would contain some members of pure lines whose average character was less than the extreme selected. The offspring, reproducing the means of their pure lines, would on the average regress back toward the mean of the population. But within a pure line, regression was complete, not partial as Galton imagined.

As for the results of continued selection in a population, Johannsen concluded:

> The usual well-known result of selection—successive progress in the direction of selection in the course of a few generations—depends . . . on the progressive purification with each generation of the deviating line concerned. And it will now be easily understood that the action of selection cannot be carried out beyond fixed limits—it must indeed cease when the purification, the isolation of the particular most strongly deviating line, practically speaking, is carried to completion.

He was hesitant, on the basis of his data, to argue that selection was absolutely ineffective in pure lines. He hedged by saying he did not claim pure lines to be "absolutely constant." But he said there was no positive evidence that selection was effective in a pure line and "the burden of proof will here rest on him who wants to assert a selection of that kind."[6]

The pure line work was significant, Johannsen believed, because it showed that fluctuating variability was not heritable, and thus was unimportant in evolution. For selection to be effective it must act upon mutations:

> The general results of this work will form an important support for the doctrine, at present especially represented by

5. Ibid., pp. 205, 206.
6. Ibid., pp. 207, 210.

Bateson and de Vries, of the great significance of "discontinuous" variation or "mutation" for the theory of heredity. For selection in populations acts in my cases only in so far as it chooses *representatives of already existing types*. These types are not successively formed . . . but they are *found* and *isolated*.[7]

Johannsen added that his pure line researches were "in full agreement with the basic ideas of the great, often mentioned works of de Vries."[8] Here indeed was new experimental evidence to support the mutation theory.

Johannsen believed his pure line investigations demonstrated that Galton's law of regression was merely a result of treating a population as an unanalyzed unit and that the law of regression did not hold for the pure lines in a population. Yet his little book was dedicated to Galton! Johannsen believed his pure line research supported "in the most beautiful manner the basic ideas of the Galtonian 'stirp' theory."[9] Continuity of the germ plasm, or stirp, meant that each individual in a pure line had the same germ plasm. Thus differences between members of the pure line were not inherited and complete regression to the type should occur each generation despite selection. Once continuity of the germ plasm was assumed, the pure line theory followed from it.

It is instructive to compare Johannsen's reasoning with Pearson's regarding Galton's law of regression. Galton himself thought that selection was ineffective in the face of regression. The deviation of the offspring of selected parents from the mean of the population was only two-thirds the deviation of the parents, and soon a balance between regression and selection must be reached. De Vries used the same argument. Pearson argued that Galton had misunderstood his own law of regression. Only if the mean of the selected parents reproduced the mean of the population would Galton's argument hold. Regression was really to the parental mean not to the mean of the population. Therefore regression did not counteract the effects of continued selection and evolution

7. Ibid., p. 212.
8. Ibid.
9. Ibid., p. 213.

could proceed continuously. Johannsen argued, like Pearson, that Galton had misunderstood his law of regression. In a pure line, regression was complete to the type of the line and selection was ineffective against such regression. Galton observed incomplete regressions because he had failed to analyze biologically his populations. If Galton had only followed the clear implications of his stirp theory, he would have understood regression from the beginning. Johannsen of course agreed with Galton's conclusion that evolution must occur in discontinuous leaps. Thus Pearson and Johannsen agreed that Galton's interpretation of his own law of regression was faulty, but they came to opposite conclusions concerning the continuity of evolution.

CRITICISM OF JOHANNSEN'S PURE LINE THEORY

Pearson and Weldon immediately published a joint reply to Johannsen's book.[10] Johannsen had claimed, on the basis of his data, that regression to the type of a pure line was complete each generation. A large bean reproduced the type of the line, as did a small bean. Thus within a pure line the correlation between parent and offspring was zero. Johannsen did not calculate the actual numerical value of parent-offspring correlation, but said his data showed it was negligible. Pearson and Weldon calculated this negligible correlation using the published data and found it was 0.3481, scarcely the zero Johannsen claimed—a powerful criticism. They also emphasized that the cumulative effects of selection might become statistically significant only after several generations of selection. In Johannsen's particular experiment, selection appeared to be effective in only one generation.

Yule also wrote a review of Johannsen's work. While believing that Johannsen had opened a fruitful new field for research, he felt the published data did not support Johannsen's conclusion that selection was ineffective within a pure line. The experiments showed only that "the effect [of selection] is small (compared with the probable error of the result)—an interesting result, but a very different matter; for if the effect

10. Karl Pearson and W. F. R. Weldon, "Inheritance in *Phaseolus vulgaris*," *Biometrika* 2 (1903): 499–503.

in given cases be not *zero* but only *small* it may in other cases be *sensible.*" [11]

The criticisms of Pearson, Weldon, and Yule were ignored amid the general acclaim which biologists accorded to Johannsen's pure line work. Here was an exciting new theory opening a new avenue of research, and it was based upon experimental evidence. Moreover, the pure line theory supported the mutation theory, which was also based upon experimental evidence. Johannsen's book of 1903 has been hailed as a very important step in the history of genetics. [12] All geneticists know that his ideas concerning heredity in pure lines were basically correct, but it is not generally known, as Pearson and Weldon pointed out, that Johannsen's data were an imperfect support for the conclusions he drew from them. The genetics literature from 1903 onward contains rare citations of the criticisms of Pearson, Weldon, and Yule; but it contains hundreds of citations of Johannsen's 1903 data as if they proved the pure line theory. The certainty was in Johannsen's mind, not in his data, as is confirmed by his next extension of the pure line research.

In the 1903 book Johannsen suggested, but did not claim, that the ineffectiveness of selection in pure lines was also true in cross-breeding or hybrid populations. In 1906 he traveled to England for the Third International Conference on Genetics where he presented a paper entitled "Does Hybridization Increase Fluctuating Variability?" [13] The conference was dominated by Mendelians. Bateson was president and chaired the meeting in which Johannsen read his paper. Still smarting from the criticisms of his pure line researches by Pearson and Weldon, Johannsen entertained the receptive audience with repeated blasts at the biometricians. Much of Johannsen's address was devoted to the defense of his earlier pure line researches. He stated that continued propagation of his pure

11. G. Udny Yule, "Professor Johannsen's Experiments in Heredity," *New Phytologist* 2 (1903): 239.

12. See L. C. Dunn, *A Short History of Genetics* (New York: McGraw-Hill, 1965), pp. 88–94.

13. In *Report of the Third International Conference on Genetics* (London: Spottiswoode, 1907), pp. 98–113.

lines showed they remained true to the type, with undiminished variability each generation. Then, on the basis of a single incomplete experiment, Johannsen extended the conclusions he had made from his pure line researches to hybrid populations.

Johannsen conducted his experiment by isolating four pure lines of *Phaseolus vulgaris* and measuring the variability in each with respect to weight and size. Then he hybridized these pure lines and measured the variability in the hybrid beans. He found that the variability was even less in the hybrids than in the pure lines; so "here was no increased amplitude of variability, offering any better material for selection." [14] Johannsen knew that in the hybrid offspring (F_2 generation)

there will be found Mendelian segregations as to dimensions and weights. This matter will be observed more closely, and the isolation of the new type-combinations shall be carried out. In this manner what may be called "unit-characters" as to length, breadth, indices, weight and so on will be elucidated. [15]

Reasoning from his faith in the stability of pure lines, Johannsen now argued that

we have no reason to suppose that an augmented fluctuation will be found in the new types which here may be formed by segregations and new combinations. Further research will, I have every conviction, give greater clearness as to the fundamental distinction of true *type differences* and *fluctuations*. [16]

Selection could of course isolate the new type-combinations found in the F_2 generation, but it was incapable of shifting the population beyond the limits of variability exhibited in the F_2 generation. Selection beyond these limits required new mutations. Thus in extending the pure line theory to hybrid populations, Johannsen reaffirmed his adherence to the mutation theory of evolution.

The basis for Johannsen's later distinction between "genotype" and "phenotype" was implicit in the distinction between

14. Ibid., p. 110.
15. Ibid.
16. Ibid.

"type differences" and "fluctuations." The phenotype was the observable organism. The genotype was the genetic constitution of the organism as determined by the factors in the gametes which united to produce the organism. In a pure line individuals had differing phenotypes but all had the same genotype; therefore selection was ineffective. In a hybrid population, however, new genotypes were formed by genetic recombination and selection could isolate these genotypes.

Johannsen's extension of the pure line theory to hybrid populations followed directly, as had the pure line theory, from the assumption of the continuity of the germ plasm. Each hereditary unit character reproduced itself exactly; unit characters could be recombined by sexual reproduction and the accompanying segregation; selection could only isolate existing recombination types; therefore, without new mutations selection was incapable of changing the population beyond the limits of variation exhibited in the F_2 generation. This was a simple, clear, and appealing generalization. Karl Pearson had used the same generalization to argue that Mendelian heredity produced insufficient variability to be an important source of variation for evolutionary change. Johannsen agreed. He thought the most important source of variation for the process of evolution came from new mutations not Mendelian recombination.

At the conference Johannsen said that in the study of heredity and evolution "what we want—in much higher degree than commonly admitted—are well analysed pure and clear elementary premises."[17] Johannsen's premises were indeed pure and clear. He expected his data to fit them. He was sufficiently certain of his reasoning to predict the results of his hybridization experiments for generations to come. But Johannsen and Pearson had overlooked a crucial possibility. They supposed that the number of recombination types was small and that each type would be represented in the F_2 generation. If, however, the number of possible recombination types was greater than they imagined, then perhaps only some of them would

17. Ibid.

appear in the F_2 generation. Others would appear for the first time in succeeding generations and might then be isolated by selection. If selection isolated a group of individuals, each with different unit characters enhancing the same trait, recombination among these characters might give rise to individuals with a genetic constitution further enhancing that trait. Thus with no new mutation, selection might progressively isolate individuals varying beyond the limits of variability exhibited in the F_2 generation.

Students of heredity found Johannsen's pure line theory and his extension of it to hybrid populations very appealing, especially in America, where de Vries's mutation theory was already popular. A new generation of biologists dedicated to experimentation had arisen in America since the generation dominated by the neo-Lamarckians Packard, Hyatt, and Cope. Members of this new generation had already repeated and confirmed de Vries's experiments with *Oenothera*.[18] Others set to work to test Johannsen's pure line theory.

HERBERT SPENCER JENNINGS AND PURE LINES

Jennings (1868–1947) worked with E. L. Mark and C. B. Davenport at the Zoological Laboratory of the Museum of Comparative Zoology at Harvard. His special interest was the behavior of lower organisms, about which he published a substantial volume in 1903. From his background of studying protozoa Jennings decided that nonconjugating lines of *Paramecium* would be ideal subjects for testing Johannsen's pure line theory. He conducted numerous experiments concerning pure lines in *Paramecium* during the years 1907–8.

In 1908 Jennings published the results of his investigations of heredity and selection in *Paramecium*.[19] Most of this long paper was devoted to a study of the effects of environment and growth upon the sizes of *Paramecium*. Both effects were found to be quite large. In the section entitled "Inheritance in

18. D. T. MacDougal, A. M. Vail, G. H. Shull, and J. K. Small, *Mutants and Hybrids of the Oenotheras,* Carnegie Institution of Washington Publication, no. 24 (Washington, D.C., 1905).

19. H. S. Jennings, "Heredity, Variation, and Evolution in Protozoa: 2. Heredity and Variation of Size and Form in Paramecium, with Studies of Growth, Environmental Action, and Selection," *Proceedings of the American Philosophical Society* 47 (1908): 393–546.

Size" Jennings described his experiments on the pure line problem. First he isolated two pure lines by choosing one individual from each of two groups whose mean sizes had remained distinct for several generations. The descendants of each individual so chosen formed a pure line. He bred these two pure lines under the same conditions for about one hundred generations, taking ten samples of progeny at different times. The mean length of one pure line varied between 112 and 162 microns, and of the other, between 86 and 106 microns. Jennings concluded that these pure lines "tend to retain the differences in size characteristic of the parents." [20] The great fluctuations in the mean length in both lines Jennings simply attributed to environmental influences. A later experiment with five pure lines under environmental conditions painstakingly equalized, extended for only twenty generations, showed that the pure lines remained distinct. But again the mean length varied up to 30 microns in one line. Jennings concluded that any fluctuations were caused by the environmental influences he had labored so diligently to eliminate and that size was strictly hereditary within any one pure line. The biometricians and a few others were unconvinced by Jennings's proof.

Having convinced himself that size was hereditary in a pure line of *Paramecium,* Jennings attempted a series of selection experiments within the pure lines. He chose large and small specimens from a pure line and found invariably that their offspring were nearly the same size: "Thus, we come uniformly to the result in all our experiments, that selection has no effect within a pure line; the size is determined by the line to which the animals belong, and individual variations among the parents have no effect on the progeny." [21] In the summary Jennings stated that "large and small representatives of the pure line produce progeny of the same mean size. The *mean size* is therefore strictly hereditary throughout the pure line." [22]

Jennings's published data on selection within a pure line,

20. Ibid., p. 487.
21. Ibid., p. 511.
22. Ibid., p. 521.

however, did not support his conclusion that the mean size was strictly hereditary. In all of his experiments he demonstrated only that the selected individuals yielded offspring of the same mean size, not that this mean size was the mean size of the pure line. The reason was probably that the mean size of the pure lines fluctuated so drastically that it was scarcely feasible to show that the offspring of the selected individuals produced the same mean size as that of the pure line as a whole.

Jennings concluded that his results were in complete agreement with those of Johannsen. Like Johannsen, he thought that selection in a population merely isolated the existing pure lines. Also like Johannsen, he later extended the same idea of the effects of selection to cross-fertilizing populations.

Soon after the publication of his technical paper on the effects of selection in pure lines of *Paramecia,* Jennings promulgated his views in more popular form in two articles. Writing as if his data had provided definite results, he claimed that "the mean length for any race is constant under given conditions" and that his pure lines were "as unyielding as iron" in the face of hundreds of generations of selection.[23] Jennings also claimed that he had isolated pure lines whose difference in mean length was five microns. His published data, to which he referred the reader, simply did not support these conclusions.

Jennings generalized the results of selection in pure lines to sexually breeding populations in a series of five propositions:

1. Organisms in which selection has shown itself effective are composed of many genotypes; of many races that are diverse in their hereditary characters. This we know to be true.
2. From such a mixture of genotypes it is possible to isolate by selection any of the things that are present—perhaps in a great number of different combinations.
3. But from such a mixture it is *not* possible to get by method-

23. H. S. Jennings, "Heredity and Variation in the Simplest Organisms," *American Naturalist* 43 (1909): 326; "Experimental Evidence on the Effectiveness of Selection," ibid., 44 (1910): 137.

ical selection anything not present (save when rare mutations have occurred).

4. Therefore it is not possible to get by methodical selection anything lying outside the extremes of the genotypic characters already existing.

5. In the case of genotypes that cross-breed readily, we may get an indefinite number of combinations of all that lies between the extremes of the existing genotypes.[24]

The first three propositions rested upon the unproved assumption that the factors in the genotype were unchangeable, except by rare mutation. Propositions four and five rested upon the unproved assumption that the possible genotypic extremes were actually existing in the population and were not merely latent possibilities. Jennings ignored the possibility of genic interaction. On the basis of these five propositions he believed he could explain almost every important selection experiment.

The results of selection in *Paramecium,* Jennings said, furnished "an excellent illustration, in the simplest possible form, of the principles of breeding for improvement so convincingly set forth in de Vries' recent work on plant breeding, and in his other writings."[25] Jennings differed from de Vries, however, in that he believed the mutations could be exceedingly small. He said the pure line work "brings out as never before the minuteness of the hereditary differences that separate the various lines."[26] Thus Jennings had found a source of variation for Darwinian selection, but the source was so meager that natural selection must require enormous amounts of time:

What the pure line work shows . . . is that the changes on which selection may act are few and far between, instead of abundant; that they are found not oftener than in one individual in ten thousand, instead of being exhibited on comparing any two specimens; that a large share of the differences between individuals are not of significance for selection or evolution. . . . Thus the work of natural selection is made infinitely more difficult and slow; but logically it is still possible.[27]

24. "Experimental Evidence" pp. 139–40.
25. Jennings, "Heredity and Variation in the Simplest Organisms," p. 331.
26. Jennings, "Experimental Evidence," p. 144.
27. Ibid.

Both Jennings and Johannsen believed they had carried the results of their pure line research to its logical conclusion. Jennings's pure line work was widely hailed as being the American corroboration of Johannsen's researches. Only a few scientists remained unconvinced.

RAYMOND PEARL AND PURE LINES

Pearl (1879–1940) received his degree at the University of Michigan and taught there as an instructor from 1902 to 1905. The year 1905/6 he spent at University College, London, working with Karl Pearson. When Weldon died in 1906, Pearson appointed Pearl as an associate editor of *Biometrika*. But Pearl soon became a Mendelian. The inevitable result was quarreling with Pearson, and in 1910 Pearl was relieved of his duties as associate editor. The years between 1907 and 1918 Pearl spent at the Maine Agricultural Experiment Station, where he conducted much research on problems in genetics.

When Pearl was converted to Mendelism he also adopted many of de Vries's ideas. Mendelism, the mutation theory, and Johannsen's pure line theory were strongly linked in his mind. When he arrived at the Maine Agricultural Experiment Station in 1907, an experiment for selection of high egg production in chickens had been in progress for nine years with no discernible success. Pearl viewed the lack of success in the selection experiment as a support for Johannsen's pure line theory. He set to work with a second experiment which he thought would relate selection for fecundity to Johannsen's work.[28]

In the first experiment (1898–1907), conducted before Pearl's arrival, from a herd of hens whose initial average egg production was 125 eggs per year, those whose production was 160 or more had been selected as the mothers for the following year. Male birds had been selected from mothers who produced 200 eggs or more per year. After nine years there was still no fixed increase in fecundity.

28. Pearl published many accounts of these experiments, some in collaboration with his assistant Frank M. Surface. Perhaps the best summary of the two initial experiments is in the article by Pearl and Surface, "Is There a Cumulative Effect of Selection?" *Zeitschrift für Induktive Abstammungs und Vererbungslehre* 2 (1909): 257–75.

In the second experiment (1907–8), conducted by Pearl and his assistant Frank M. Surface, the females were of two classes: birds laying 160–99 eggs in their pullet year and birds laying 200 or more. Cockerels from mothers who had laid more than 200 eggs in their pullet year were the male parents for both groups. The result was that the mothers with lower egg production (160–99 per year) yielded offspring with higher fecundity than the offspring of the more productive mothers (200 or more eggs per year). Selection was ineffective in raising the level of egg production. Indeed, in this experiment, the correlation between mother and daughter with respect to egg production was negative.

On the basis of these two experiments, Pearl concluded, "So far as the character fecundity (egg production) in the domestic fowl is concerned long continued and carefully executed experiments give no evidence whatever that there is a cumulative effect of the selection of fluctuating variations." [29] Pearl believed this result supported the conclusions drawn by Johannsen and Jennings from their work in pure lines.

The experiments of Jennings and Pearl were widely quoted as supporting Johannsen's pure line theory. By 1910 the pure line theory and the selection theory associated with it were generally accepted in both Europe and America. The evidence seemed adequate to most geneticists, but Karl Pearson and J. Arthur Harris, an American biometrician, had some pointed criticisms.

CRITICISM OF THE PURE LINE THEORY

Pearson had already found discrepancies between Johannsen's data of 1903 and his conclusions based upon them. In 1910 he critically analysed the pure line experiments conducted since 1903.[30] Most of his paper was devoted to an analysis of the data of Elise Hanel, who had conducted selection experiments on pure lines of *Hydra grisea*.[31] Jennings had repeatedly cited Hanel's work as further experimental support for Jo-

29. Ibid., p. 272.
30. Karl Pearson, "Darwinism, Biometry, and Some Recent Biology," *Biometrika* 7 (1910): 368–85.
31. Elise Hanel, "Vererbung bei Ungeschlechtlicher Fortpflanzung von Hydra Grisea," *Jenaische Zeitschrift* 43 (1908): 322–72.

hannsen's theory. Pearson demonstrated that her data did not yield the decisive support for the pure line theory that she and Jennings claimed.

The fundamental error in the experiments of Jennings and Pearl seemed obvious to Pearson. He saw that neither researcher had adequately demonstrated the heritability of the characters in question. Jennings had stated on the basis of his data that mean size was "strictly hereditary" in a pure line and that his pure lines were "as unyielding as iron." Pearson thought that Jennings's data did not warrant such conclusions. He argued that Jennings "assumes that by selecting a character, the heredity of which he has never demonstrated, he can reach a general and 'absolute demonstration' of the truth of the theory of pure lines!"[32] The same criticism applied to Pearl's work. Pearl had never demonstrated that fecundity was inherited.

Pearl's belief that his experiments with fecundity supported Johannsen's pure line theory is scarcely defensible. Johannsen said selection was ineffective because it was acting upon identical genotypes. Pearl was well aware that his populations of chickens were genetically diverse, even in the genes controlling fecundity. Therefore the reason for the failure of selection in Pearl's experiments was different from the reason in Johannsen's experiments. Johannsen maintained that selection was effective in genetically diverse populations up to certain limits. If selection were completely ineffective, he thought the population must be composed of like genotypes. This was clearly not the case in Pearl's chickens.

Jennings answered Pearson's criticism with some ugly remarks. He asked, "Are there any biologists of achievement that still hold with Pearson?"[33] He also claimed Pearson was so unreasoning in biological matters that to be attacked by him was an honor:

Those who find the genotype idea useful may . . . prepare themselves for one of those justly famous bludgeonings from

32. Pearson, "Darwinism, Biometry, and Some Recent Biology," p. 373.
33. Jennings, "Experimental Evidence," p. 143.

the dictator of the whilom orthodox biometrical school; this is the last honorable mark of distinction which stamps the investigator as a thorough and exact analyst of things biological.[34]

Jennings did not bother to answer Pearson's scientific criticism. But he soon started a new series of pure line experiments which were to show that Pearson's criticism carried weight. Pearl reacted to Pearson's criticism by increasing his efforts to discover just how fecundity was inherited.

J. Arthur Harris, with Davenport and Pearl, was one of the leading biometricians in America. He was not opposed to Johannsen's pure line theory, as was Pearson, but he did think the experimental proof of the pure line theory was weak. Harris delivered his criticisms at a symposium on "The Study of Pure Lines of Genotypes" before the American Society of Naturalists on 29 December 1910. As an introduction Harris said:

> On this platform I find myself in a somewhat embarrassing position. A friend assured me in advance that this symposium would be somewhat analogous to the country parson's "praise service," and into this pure devotional atmosphere I must bring a note of agnosticism.[35]

He went on to say:

> Our symposium has for its subject the Genotype or Pure Line Theory. Some of the speakers have enthusiastically urged us to replace the words "pure line theory" by "pure line facts." If this were done there would be little need for this program. Pure line facts are as yet a very insignificant part of biological data. The real occasion for this symposium is the pure line theory—the rank vines which have grown from the nineteen bean seeds which Johannsen planted in 1901.[36]

Harris reviewed the major experiments which supposedly proved the pure line theory. He concluded that although he was receptive to the theory, the data did not warrant the conclusions which had been drawn from them. The heart of

34. Ibid.
35. J. Arthur Harris, "The Biometric Proof of the Pure Line Theory," *American Naturalist* 45 (1911): 346.
36. Ibid., p. 347.

Harris's criticisms was that the upholders of the pure line theory were guilty of circular reasoning: (1) they assumed continuity of the germ plasm; (2) they observed that selection could not change the pure line (if it did, then the material was impure); (3) they concluded that selection in pure lines was impossible. The conclusion of course followed directly from the initial assumption, as did the observation. Harris thought Johannsen especially was guilty of this kind of reasoning.

Harris accurately pinpointed the real reason why the pure line theory was accepted so readily on the basis of faulty scientific evidence. The climate of biological opinion was favorable for the pure line theory. Many biologists simply assumed the continuity of the germ plasm, and the pure line theory followed from this. Johannsen's decision to dedicate his original pure line work to Galton because of the "stirp" theory was entirely fitting.

Johannsen, Jennings, and Pearl all concluded on the basis of pure line research that selection in sexually breeding populations could produce nothing beyond the normal range of variability, unless new mutations occurred. In 1910 the pure line theory seemed so obvious that most outstanding geneticists accepted it without adequate proof. Most of them also accepted the related selection theory, and the two ideas became firmly associated.

Jennings concluded that evolution probably proceeded by small steps in an enormously protracted process. Johannsen and Pearl stuck to de Vries's mutation theory. In 1910 the pure line evidence appeared to demonstrate that selection was an ineffective process in evolution. Ten years later most outstanding geneticists rejected this conclusion.

The Proof and Explanation of the Effectiveness of Selection

Only a few Mendelians in the years 1901–11 believed that Darwinian selection was the most important factor in species change; most believed in discontinuous evolution. Yet by 1918 the effectiveness of Darwinian selection was widely recognized by Mendelians. Three essential developments before 1912 brought about this change of view: (1) William Castle's experiments which demonstrated that selection of continuous variations was effective in changing a character to a new stable level beyond the original limits of variation of the character; (2) the experimental demonstration by H. Nilsson-Ehle and Edward East that Mendelism could explain some continuous variations and what appeared to be blending inheritance; (3) the discovery by T. H. Morgan and the *Drosophila* workers that Mendelian characters might be very small variations. These three experimental results were found in the years 1908–11, the same period when Johannsen, Jennings, and Pearl were pointing out difficulties of the Darwinian selection theory. The years 1912–18 witnessed the resolution of most of the remaining inconsistencies between Darwinian selection and Mendelism.

WILLIAM ERNEST CASTLE AND SELECTION THEORY

Castle (1867–1962) studied with C. B. Davenport at Harvard, where he received his Ph. D. in 1895. His researches up to 1900 concerned embryology and developmental problems. During the years 1900–1902, a period in which his scientific interests were changing and he was not publishing, he became acquainted with Mendelian heredity. In January 1903 he published a paper describing Mendelian inheritance, calling it "one of the great discoveries in biology, and in the study of heredity perhaps the greatest." [37] He published five other papers on Mendelian heredity that year.

Castle was much influenced by reading Bateson's fiery *Mendel's Principles of Heredity: A Defence.* He immediately supported Bateson in opposition to the biometricians. Like Bateson, he was not trained in mathematics and was hopelessly confused by Pearson's revisions of the law of ancestral heredity. Unlike Bateson, Castle lacked the good sense to avoid a confrontation with the biometricians on their own ground. In 1903 he published a paper entitled "The Laws of Heredity of Galton and Mendel, and Some Laws Governing

37. W. E. Castle, "Mendel's Law of Heredity," *Science* 18 (1903): 396–97.

Race Improvement by Selection." In this paper he examined G. von Guaita's data on inheritance in mice and concluded that "some fundamental defect exists in 'the law of ancestral heredity,' as stated by either Galton or Pearson."[38] The problem was that Castle did not understand the differences between the ancestral correlation coefficients, the multiple regression coefficients, and the percentages of offspring resembling an ancestor with regard to a particular character.

Pearson replied with a scathing note in *Biometrika*. He demonstrated Castle's lack of understanding concerning the law of ancestral heredity and concluded that "either Professor Castle is so ignorant that he does not know that a coefficient of correlation cannot be a group frequency; or, he has directly misquoted my memoirs because any form of argument suffices for the audience he wishes to appeal to."[39] Pearson then issued a direct challenge to Castle either to prove mathematically that the law of ancestral heredity (as interpreted by Pearson) did not apply to von Guaita's data on mice or to retract his statements. Castle and Pearson were never afterward on friendly terms, but curiously, within five years Castle had adopted two of Pearson's most basic positions against the Mendelians: that the selection of small differences was effective as a means of species change and that selection should be effective within a pure line.

Probably because he was influenced by reading Bateson's works, Castle at first accepted the seemingly obvious connection between Mendelism and discontinuous evolution. On 28 December 1905 he attended a meeting of the American Society of Naturalists devoted to the mutation theory of evolution. Joined by E. G. Conklin, William Morton Wheeler, D. T. MacDougal, and Liberty Hyde Bailey, Castle read a paper advocating the mutation theory. He began by saying that Darwin was right to recognize "that there is no essential

38. *Proceedings of the American Academy of Arts and Sciences* 39 (1903): 226.

39. Karl Pearson, "A Mendelian's View of the Law of Ancestral Inheritance," *Biometrika* 3 (1904): 110.

difference between breeds and species, and that if we can ascertain how breeds originate we can infer much as to the origin of species." [40] Current evidence indicated "that the material used by breeders for the formation of new breeds consists almost exclusively of mutations," [41] a position which Darwin had rejected. Castle concluded that the analogy between artificial and natural selection led one to accept the mutation theory.

He cited an example from his own experience. In 1901 a guinea pig with a supernumerary fourth digit on one of its hind feet appeared in his stocks. After five generations of selection for the added digit he was able to create a new race of polydactylous quinea pigs. "This race was not *created* by selection, though it was *improved* by that means." [42] The overwhelming impression left upon Castle's mind by this mutation in his stocks, and by several others he reported, was that the mutation theory was correct and that the selection of continuous variations was ineffective. "Modification of characters by selection, when sharply alternative conditions (*i.e.,* mutations) are *not* present in the stock, is an exceedingly difficult and slow process, and its results of questionable permanency." [43]

In the light of his later research Castle's paper favoring the mutation theory seems curiously out of character. In his book *Heredity in Relation to Evolution and Animal Breeding,* published in 1911, Castle reevaluated the origin of his race of polydactylous guinea pigs. Now his emphasis was on the process of selection not on the original mutation:

I have observed characters at first feebly manifested gradually improve under selection until they become established racial traits. Thus the extra toe of polydactylous guinea-pigs made its appearance as a poorly developed fourth toe on the left foot only. . . . Individuals were selected throughout five

40. W. E. Castle, "The Mutation Theory of Organic Evolution from the Standpoint of Animal Breeding," *Science,* n. s. 21 (1905): 522.
41. Ibid.
42. Ibid., p. 523.
43. Ibid., p. 524.

successive generations, at the end of which time a good four-toed race had been established.[44]

What had caused this reversal of Castle's position?

The change began soon after Castle's address advocating the mutation theory. In June 1906, Hansford MacCurdy, under Castle's direction, completed a study of the inheritance of coat color in rats. He found that the piebald pattern of "hooded" rats behaved as a Mendelian recessive to the gray color of wild rats. MacCurdy and Castle carried out two important experiments with the hooded rats. First, they crossed hooded rats with rats bearing the "Irish" pattern. The "Irish" pattern was like the hooded pattern except that instead of a black dorsal stripe the entire back and sides of the rat were black. The hybrids were then back-crossed with the parental hooded stock to obtain offspring with the hooded pattern again. The average size of the dorsal stripe in the hooded pattern of these rats was considerably raised. Second, they selected for increased and decreased size of the dorsal stripe in the hooded rats. Selection was carried on for five generations with definite success in both directions and with no indication that regression would move either selected group back to its original characteristics.

On the basis of these experiments Castle and MacCurdy challenged the view of evolution proposed by de Vries. De Vries, they said, concluded that continuous variations were not inherited at all, except temporarily, and that selection of continuous variations produced only temporary modifications which speedily disappeared when selection ceased. "This conclusion, however, seems to us altogether too sweeping. They described the basic problem:

> De Vries maintains that all species-forming variations are of this sort [mutations]; that selection is unable to form new species, because it can neither call into existence mutations nor permanently modify a race by cumulation of abmodal fluctuations. Darwin, on the other hand, and the great majority of his followers, while admitting that races are occasionally produced

44. W. E. Castle, *Heredity in Relation to Evolution and Animal Breeding* (New York: D. Appleton, 1911), pp. 120–21.

by discontinuous or "sport" variation, ascribe evolutionary progress chiefly to the cumulation through long periods of time of slight individual differences, such as de Vries calls fluctuations. The issue between the two views is sharp and clear. According to de Vries, if we rightly understand him, selection is not a factor in the *production* of new species, but only in their *perpetuation,* since it determines merely what species shall survive; according to the Darwinian view, new species arise through the direct agency of selection, which leads to the cumulation of fluctuating variations of a particular sort.

Castle and MacCurdy concluded that their experiments supported "the Darwinian view rather than that of de Vries." [45]

They knew their selection experiment on hooded rats was effective in changing the shape of the hooded pattern—a fact that supported Darwin's theory. Their explanation for the success of the selection experiment proved, however, to be controversial. From the hybridization experiment they concluded that

though the inheritance is clearly Mendelian, when hooded and Irish rats are crossed, the gametes formed by cross-breds are not pure, but modified, each extracted pattern being changed somewhat in the direction of that pattern with which it was associated in the cross-bred parent. This means simply that the inheritance, though in the main alternative, is to some extent blending.[46]

Castle and MacCurdy reasoned that selection also changed the Mendelian factor for the hooded pattern. Some influential biologists, among them C. B. Davenport and T. H. Morgan, at this time agreed with Castle that the Mendelian factors could be altered. But one effect of Johannsen's pure line theory was to convince many Mendelians that the hereditary factors were very stable. Castle's conclusion concerning the modifiability of Mendelian factors was certain to be controversial. At the same time that Johannsen, Jennings, and Pearl

45. W. E. Castle and Hansford MacCurdy, *Selection and Cross-Breeding in Relation to the Inheritance of Coat-Pigments and Coat-Patterns in Rats and Guinea-Pigs,* Carnegie Institution of Washington Publication, no. 70 (Washington, D.C., 1907), pp. 2, 3, 4.
46. Ibid., p. 34.

were claiming on the basis of pure line theory that selection was incapable of changing a character beyond the original limits of variation unless accompanied by new mutation, Castle was working on selection experiments to prove them wrong.

In October 1907, Castle initiated a new series of selection experiments with hooded rats. With several associates he continued these experiments until 1919. Approximately fifty thousand rats were bred in these years. When Castle reported on the progress of the experiments in 1911, the success of selection for both increased and decreased size of the hooded pattern was marked:

> The interesting feature of this experiment is the production, as a result of selection, of wholly new grades; in the narrow series, of animals having less pigment than any known type other than the albino; in the wide series, of animals so extensively pigmented that they would readily pass for the "Irish type," which has white on the belly only, but which is known to be in crosses a Mendelian alternative to the hooded type. By selection we have practically obliterated the gap which originally separated these types, though selected animals still give regression toward the respective types from which they came. But this regression grows less with each successive selection and ultimately should vanish. . . . As yet there is no indication that a limit to the effects of selection has been reached.[47]

Thus by 1911 Castle's experiments had contradicted the assertions of the pure linists concerning cross-breeding populations. Castle had been able to produce new types by selection. In the coming years he was to champion the selection theory.

THE MULTIPLE FACTOR THEORY

Johannsen had shown in 1903 why selection was effective up to the existing limits of variation in a population of pure lines. Selection merely isolated the pure lines which varied farthest in the direction of selection. But the situation was more complex in the case of cross-breeding populations, unless the entire population was homozygous for the characters in question. There were no pure lines in Johannsen's sense.

47. Castle, *Heredity*, pp. 125–26.

How could the heritability of some apparently continuous variations, as in Castle's experiments, be explained? Mendel, on the basis of experiments with flower color in *Phaseolus,* had suggested that if two or more independent factors were involved, an apparently continuous array of variations might be the result. Yule outlined this possibility again in 1902, and Pearson in 1903. Both Bateson and Hurst considered multiple factor inheritance to be a probable explanation of apparently continuous variations.[48] The concept of multiple factor inheritance was known in the early 1900s. In 1908 H. Nilsson-Ehle supplied an experimental proof.

In 1900 Nilsson-Ehle became an assistant to Hjalmar Nilsson, director of the Swedish Agricultural Experiment Station at Svalöf. His specialty was breeding cereals, and in the years 1900–1908 he carried out numerous crosses with various varieties of oats and wheat. He published a short account of his work in 1908, and in 1909 he published a 122-page monograph.[49]

Nilsson-Ehle was acquainted with Mendel's account of experiments with *Phaseolus;* so he was prepared to interpret the unusual ratios which appeared in some of his crosses. Many of the characters treated in both oats and wheat revealed the usual 3:1 ratio in the F_2 generation. But other very different ratios appeared. Nilsson-Ehle made four crossings of two varieties of oats, one with black glumes and the other with white. He discovered an average segregation ratio of 15.8 black:1 white in the F_2 generation,[50] which was very close to the 15:1 ratio predicted by Mendel for a two-factor segregation. In the F_3 generation, if the Mendelian hypothesis were true, the black parents (from F_2) should produce a ratio of 7 (producing only black):4(segregating 3 black:1 white)

48. In a letter dated 24 March 1902, Bateson told Hurst, "I have no doubt you are quite right in believing that Mendelian principles may apply as well to blended as to alternative inheritance" (BPB 21).

49. H. Nilsson-Ehle, "Einige Ergebnisse von Kreuzungen bei Hafer und Weizen," *Botaniska Notiser* (1908), pp. 257–94; "Kreuzungsuntersuchungen an Hafer und Weizen," *Lunds Universitets Årsskrift,* n.s., ser. 2, vol. 5, no. 2 (1909).

50. "Kreuzungsuntersuchungen," p. 29.

:4(segregating 15 black:1 white). The observed ratio was
7:4.5:3.7. The F_3 generation consisted of only 39 plants, and
Nilsson-Ehle concluded that "one could scarcely expect a better
agreement with such a small number of individuals." [51]

In experiments with winter wheat, Nilsson-Ehle found another two-factor segregation in the chaff color. In four different crossings of the same varieties of brown and white
chaffed wheat, he obtained in the F_2 generation a total of
1,410 brown, light-brown:94 white—exactly 15:1. [52] The brown
color exhibited incomplete dominance in this experiment and
many shades of brown appeared in the F_2 generation.

His most striking results came in crossing wheats of different grain color. Most such crosses showed the usual 3:1 F_2
ratio. But one red-grained Swedish velvet wheat produced
different results. Nilsson-Ehle made five crosses of the red-
grained wheat with a white variety; all F_1 individuals obtained were red, but lighter red than the parents. Each of
these five matings produced only red offspring in the F_2. Un-
dismayed, he took the 78 F_2 kernels from one of the matings
and planted them. Fifty of the kernels yielded only red offspring; five segregated approximately 63 red:1 white; fifteen
segregated 15 red:1 white; and eight segregated 3 red:1
white. The results suggested that the kernel color was determined by three independent factors. Nilsson-Ehle weighted
his observed results by supposing that five of the kernels
yielding only red F_3 offspring were really segregating 63 red:
1 white. Then he calculated the expected F_3 ratio on the
three-factor hypothesis and compared the weighted observed
ratio; his results [53] are given in table 1. Since there were no
F_2 white kernels to breed true, Nilsson-Ehle believed his data
furnished excellent support for the three-factor Mendelian
interpretation.

The immense possibilities for genetic combinations caused
by Mendelian segregation impressed Nilsson-Ehle. He was
aware that his researches had covered only the easiest cases,

51. Ibid., p. 32. The quotes from Nilsson-Ehle are my translations.
52. Ibid., p. 61.
53. Ibid., p. 70.

TABLE 1
Nilsson-Ehle's Three-factor Segregation in Red-grained Wheat

	Constant Red	Segregating			Constant White
		63:1	15:1	3:1	
Expected	37	8	12	6	1
Observed	37	8	12.3	6.6	0

with one, two, or three segregating factors. He did find one case which suggested that four segregating factors controlled the presence of a ligule in oats, but the evidence was inconclusive.[54] But what if there were many independent segregating factors? With incomplete dominance, which Nilsson-Ehle had observed in many of his cases, ten independent factors provided "nearly 60,000 different possible forms," [55] each with a different genotype.

On the basis of his understanding of genetic recombination, Nilsson-Ehle made two crucial conclusions:

1. It is probable "that many mutations, above all in exotic plants, are only new groupings of already present factors and really represent nothing new, especially in such cases where they throwback," [56] that is, in cases of atavism. Populations were often sufficiently small in comparison with the number of possible genetic recombinations that some combinations would appear only rarely; but these were not cases of new mutation. Atavism had caused Darwin and others much consternation. Here was experimental indication of a scientific explanation.

2. The primary purpose of sexual reproduction must be to increase the possibility of genetic recombinations. Natural selection operating upon these organisms with differing genotypes enables a population to adapt to changing environments.

54. Ibid., pp. 75–91.
55. Ibid., p. 116. The number of different possible genotypes with n independent loci and two allelles at each locus is 3^n. With incomplete dominance each genotype is phenotypically distinguishable. Thus in Nilsson-Ehle's example there are $3^{10} = 59,049$ types. With complete dominance the expected number of distinguishable phenotypes is 2^n.
56. Ibid., p. 109.

Nilsson-Ehle provided experimental verification of a source of apparently continuous variation upon which natural selection could act. The multiple factor interpretation of continuous variation was as easily accepted as the pure line theory, and for the same reason. If selection of continuous variations was successful, there had to be an explanation consistent with the continuity of the germ plasm. Nilsson-Ehle emphasized that his researches were consistent with "the purity of the gametes in Mendel's sense." [57]

Nilsson-Ehle's experimental discoveries soon became widely known. The journals of heredity and every major genetics textbook after 1910 referred to his work. Although American geneticists knew about Nilsson-Ehle's work, most of them apparently read secondary accounts. The original article, written in German, was printed in a Swedish journal carried by few libraries. The consequence was that in America the multiple factor theory quickly gained acceptance, but the implications of sexual reproduction as a source of variation for evolution were not widely understood by American geneticists for almost ten years following the publication of Nilsson-Ehle's important paper.

In the years 1910–18 Edward Murray East (1879–1938) performed the same service in America that Nilsson-Ehle had on the Continent. He demonstrated experimentally that Mendelian inheritance could account for an almost continuous array of variation, and he elucidated the role of sexual reproduction in evolution. East was a graduate student at the University of Illinois between 1900 and 1905. Trained as a chemist, he worked at the Illinois Agricultural Experiment Station with C. G. Hopkins on selection experiments to change the oil and protein content of corn. Hopkins began the experiments in 1898 and was selecting for high and low oil and protein content. East's job was to chemically analyse the corn samples. He was impressed by the success of the selection experiment and he worked in the following years toward an understanding of how small heritable variations

57. Ibid., p. 118.

were produced in cross-breeding organisms. In October 1905 East accepted an appointment to undertake plant breeding at the Connecticut Agricultural Experiment Station. Here he conducted numerous experiments on inbreeding and cross-breeding in maize, potatoes, and tobacco.

By the time East accepted a position at the Bussey Institution of Harvard University in 1909, his investigations with maize had led him to believe in a multiple factor interpretation of continuous variation. At this time he became acquainted with Nilsson-Ehle's 1909 paper and he decided to publish the results of his own preliminary investigations. In a paper entitled "A Mendelian Interpretation of Variation That Is Apparently Continuous," [58] East presented the basic arguments for multiple factor inheritance accompanied by evidence from his experiments with maize. In 1910, however, he apparently had not discovered the full implications of genetic recombination for evolutionary adaptation. Discussing the Illinois corn selection experiment he stated: "It is very evident that the original stock was a mixed race containing sub-races of various composition intermingled by hybridization. Selection rapidly isolated these sub-races. . . . After this selection accomplished nothing." [59] He also said the Illinois selection experiment "has given a complete corroboration of Johannsen's conclusions on pure lines." [60] East was unaware at this time of the immense possibility for genetic recombination to occur within the population and was in substantial agreement with Johannsen, Jennings, and Pearl. But East and Castle were friends at the Bussey Institution, and soon Castle had selected beyond the former limits of variation in hooded rats. Furthermore, contrary to East's prediction, selection continued to be effective in the Illinois corn selection experiment. East suggested in 1910 that selection applied after hybridization might be effective in changing a character beyond its

58. Edward M. East, "A Mendelian Interpretation of Variation That is Apparently Continuous," *American Naturalist* 44 (1910): 65–82.
59. Edward M. East, "The Role of Selection in Plant Breeding," *Popular Science Monthly* 77 (1910): 199.
60. Ibid., p. 198.

former limits of variation.[61] He was later to extend his ideas on multiple factor inheritance to cover cases of selection in a single cross-breeding population, as Nilsson-Ehle had done in 1909.

THOMAS HUNT MORGAN AND VARIATION FOR EVOLUTION

Another possible source of small hereditary variations was discovered by Morgan (1866–1945) and his students. In his book *Evolution and Adaptation,* published in 1903, Morgan severely criticized the adequacy of Darwin's idea of natural selection as the mechanism for the process of evolution. Between 1900 and 1909 he was also critical of Mendelian inheritance, but by 1909 his views on evolution had begun to change. In an article entitled "For Darwin" he suggested that some continuous variations were indeed inherited:

> We have discovered . . . that *some small* variations are inherited. Let us call these *definite variations,* and if these be the material with which evolution is concerned, Darwin's assumption in regard to the nature of variation will be, in part, justified.[62]

Morgan's analysis of the possible role of these "definite variations" in evolution was curious. If a new definite variation appeared in a population, then gradually

> step by step, the new character will be added to the whole race. Thus any new, definite character will gradually appear in all the individuals whether it is useful or not. If it is useful it may sooner implant itself on the race than if it is indifferent; for more individuals may survive that possess it, than of those without it. It will spread faster, but in any case it will come in the long run. Thus we see that it spreads, not because it is advantageous, but because it is a definite variation.[63]

Morgan obviously was unaware of the Hardy-Weinberg equilibrium in 1909. He even neglected to take segregation into account in his theory of evolution by small definite variations. But he was warming up to the view that evolution depended upon small variations.

61. East, "A Mendelian Interpretation of Variation," p. 81.
62. T. H. Morgan, "For Darwin," *Popular Science Monthly* 74 (1909): 375.
63. Ibid., p. 377.

Morgan's conversion to Mendelism in 1910 as a consequence of his *Drosophila* experiments is well known.[64] He and his co-workers discovered very small Mendelian characters in *Drosophila* by 1912. Here were the unswampable, heritable, small variations which made Darwin's idea of evolution plausible to Morgan.

By 1912 each of the three developments in selection theory traced above were well known to geneticists. Castle had selected hooded rats beyond their former limits of variation. Nilsson-Ehle and East had demonstrated the Mendelian basis of some phenotypically continuous variations. The *Drosophila* workers had demonstrated that Mendelian variations could be very small. This work brought Mendelism and Darwinism closer, but many issues still remained to be settled before the gulf between them would melt away: (1) Were de Vries's "mutations" in *Oenothera* really mutations? (2) What were the implications of the pure line work for selection theory? (3) What was the role of sexual reproduction in evolution? (4) What were the implications of the *Drosophila* work for evolutionary theory? (5) Did Castle's selection experiments show that Mendelian factors could be changed?

OENOTHERA MUTANTS

Because of the regularity with which mutant forms appeared in de Vries's *Oenotheras,* Bateson and others suspected that some of the *Oenotheras* were permanent hybrids. The experimental work of B. M. Davis between 1909 and 1916 lent support to this view. Summarizing the evidence in 1916, Castle concluded:

Oenothera Lamarckiana is best interpreted as an impure or hybrid species which only breeds true in a relatively high degree because of extensive sterility, which eliminates large numbers of gametes and zygotes that differ from the germinal cells which reproduce the *Lamarckiana* type. The "mutants" come from occasional seeds of different types that survive the heavy mortality which renders sixty per cent or more of the seeds infertile and about fifty per cent of the pollen

64. For a good account, see A. H. Sturtevant, "Thomas Hunt Morgan, 1866–1945," *Biographical Memoirs of the National Academy of Sciences* 33 (1959): 283–325.

grains abortive. If this is the correct explanation of the peculiar breeding behavior of *Lamarckiana,* this plant is very far from being representative of a pure species, as de Vries assumed it to be, and is hardly suitable material for experiments designed to give evidence of mutation.[65]

In 1918 H. J. Muller suggested a forceful interpretation of the behavior of the *Oenotheras.* In a paper which treated balanced lethals in *Drosophila,* he noted the resemblance to the *Oenotheras* in the appearance of offspring: "It is difficult in view of all the parallelisms to believe that the two sets of phenomena have not a similar basis, and that the Oenotheras do not represent a complicated case of balanced lethals." [66] The proof of Muller's hypothesis had already been demonstrated for *Oenothera Lamarckiana* by Renner in 1917, but his work was not generally known until later.[67] The reliability of mutations in *Oenothera* as a proof of the mutation theory of evolution was severely shaken by 1918 even though the actual genetic mechanisms were not generally known. Because de Vries's major proof had come from the *Oenotheras,* skepticism about his mutation theory increased.

PURE LINE THEORY AND SELECTION

In 1911 both Jennings and Pearl had declared on the basis of pure line theory that selection in cross-breeding populations was incapable of changing a character beyond the existing limits of variation. In this conclusion they agreed with Johannsen. Pearson, Harris, and Castle attacked the pure line work as inconclusive. In response Jennings initiated a new series of experiments with a shelled rhizopod, *Difflugia corona* and found that selection was capable of changing his pure lines. When his results were published in 1916, he no longer lashed out at his former critics:

The criticisms of negative results [of selection in pure lines] as due to the fact that the characters worked with are largely

65. W. E. Castle, *Genetics and Eugenics* (Cambridge: Harvard University Press, 1916), p. 81.
66. H. J. Muller, "Genetic Variability, Twin Hybrids and Constant Hybrids, in a Case of Balanced Lethal Factors," *Genetics* 2 (1918): 471.
67. For an account of Renner's work, see A. H. Sturtevant, *A History of Genetics* (New York: Harper and Row, 1965), pp. 62–66.

the expression of the particular growth stage of the organism, and its environment up to the time studied, take on much weight when one sees, on the one hand, how long it may require for selection to give an inherited effect even with congenital characters; on the other how extremely marked the results in time become with such congenital characters.[68]

Jennings suggested that a reason for the success of selection in these experiments might be undetected genetic recombination in *Difflugia*. Thus at this time Jennings believed that genetic recombination could provide more heritable variability than he had thought in 1910. As a consequence he placed more emphasis on the efficacy of selection.

In 1917 Jennings wrote that Mendelian recombination and Darwinian selection, taken together, were the keys to understanding evolution:

It appears to me that the work in Mendelism, and particularly the work on *Drosophila,* is supplying a complete foundation for evolution through the accumulation by selection of minute gradations. We have got far away from the old notion that hereditary changes . . . are bound to occur in large steps. The "multiple allelomorphs" show that a single unit factor may exist in a great number of grades; the "multiple modifying factors" show that a visible character may be modified in the finest gradations by alterations in diverse parts of the germinal material. The objections raised by the mutationists to gradual change through selection are breaking down as a result of the thoroughness of the mutationists' own studies.

The positive contribution of these matters to the selection problem is to enable us to see the important role played by Mendelism in the effectiveness of selection. . . . By selective cross-breeding it is possible to bring together into one stock all the modifiers that have been produced in diverse stocks. Mendelism acts as a tremendous accelerator to the effectiveness of selection.[69]

Jennings now believed fully in the efficacy of Darwinian selection.

68. H. S. Jennings, "Heredity, Variation, and the Results of Selection in the Uniparental Reproduction of *Difflugia Corona," Genetics* 1 (1916): 523.

69. H. S. Jennings, "Modifying Factors and Multiple Allelomorphs in Relation to the Results of Selection," *American Naturalist* 51 (1917): 305–6.

Unlike Jennings, Pearl stuck by his earlier conclusions about selection based on pure line theory. In 1917 he stated that on the basis of current evidence "the conclusion will be reached that natural selection is no longer generally regarded as the primary, or perhaps even a major, factor in evolution."[70] He argued that somatic fluctuations were in general so large that selection, which operates upon somatic differences, could never assure that selected individuals would leave their characters to their offspring. Besides, said Pearl, "observation indicates that in many cases evolutionary changes have come about by relatively large, discontinuous leaps."[71]

Pearl's argument that no significant correlation existed between genotypes and phenotypes must have seemed strange even to his contemporaries. In 1917 Pearl was moving away from the mainstream of evolutionary thought, despite his statement to the contrary. He was, nevertheless, part of a significant minority. Fisher, Haldane, and Wright each suggested that his work was in part a reaction to Mendelians who still believed in discontinuous evolution.

THE ROLE OF SEXUAL REPRODUCTION IN EVOLUTION

Nilsson-Ehle clarified in 1909 the importance of sexual reproduction in evolution: genetic recombination furnished an immense amount of heritable variation. This seems to have been realized more slowly in England and America. During the years 1912–18 English and American geneticists gradually recognized the importance of the variation produced by recombination, but in 1918 East still felt the need to publicize the point. In an article entitled "The Role of Reproduction in Evolution" he emphasized the vast amount of heritable variation created by genetic recombination:

If N variations occur in the germplasm of an asexually reproducing organism only N types can be formed to offer raw material to selective agencies. But if N variations occur in the germplasm of a sexually reproducing organism 2^N types can be formed. The advantage is almost incalculable.[72]

70. Raymond Pearl, "The Selection Problem," ibid., p. 73.
71. Ibid.
72. Edward M. East, "The Role of Reproduction in Evolution," *American Naturalist* 52 (1918): 284. East was obviously speaking of the case in which dominance was complete.

East concluded that "the essential feature of the role of repro-duction in evolution is the persistence of mechanisms in both the animal and plant kingdoms which offer selective agencies the greatest amount of raw material." [73] If Nilsson-Ehle's im-portant paper of 1909 had been translated into English and made readily available, the significance of sexual reproduc-tion as a source of hereditary variation might have been gen-erally recognized sooner.

MORGAN'S THEORY OF EVOLUTION

When Morgan grasped the meaning of Mendelian segrega-tion from the *Drosophila* work, he came to understand that a new mutation would not automatically spread through a pop-ulation. With the encouragement and prodding of his co-workers, H. J. Muller, A. H. Sturtevant, and C. B. Bridges, Morgan came to believe in the importance of selection in evolution. His idea of evolution was simple and appealing, a natural outgrowth of the *Drosophila* work. In *A Critique of the Theory of Evolution,* containing lectures he delivered in February and March 1916, Morgan stated his views concern-ing mutation and selection:

> If through a mutation a character appears that is neither ad-vantageous nor disadvantageous, but indifferent, the chance that it may become established in the race, i.e., as a racial characteristic is extremely small, although by good luck such a thing may occur rarely. . . . If through a mutation a char-acter appears that has an *injurious* effect, however slight this may be, it has practically no chance of becoming established.
>
> If through a mutation a character appears that has a *benefi-cial* influence on the individual, the chance that the individ-ual will survive is increased, not only for itself, but for all of its descendants that come to inherit this character. It is this increase in the number of individuals possessing a particular character, that might have an influence on the course of evo-lution. [74]

Ernst Mayr has characterized this view as the "beanbag" the-ory of evolution. [75] It is a theory of single gene replacements.

73. Ibid., p. 289.
74. T. H. Morgan, *A Critique of the Theory of Evolution* (Prince-ton: Princeton University Press, 1916), 187–90.
75. Ernst Mayr, "Where Are We?" *Cold Spring Harbor Symposia on Quantitative Biology* 24 (1959): 1–14.

Morgan's theory was consistent with Darwinian evolution. The *Drosophila* work demonstrated what Jennings and Nilsson-Ehle had already claimed: that heritable variations might be very small. The simplicity of Morgan's view appealed to geneticists and the "beanbag" theory gained widespread approval. Fisher, Haldane, and Wright, however, were to disagree, about the importance of Morgan's concept of evolution.

CASTLE AND THE SELECTION PROBLEM

In 1911 Castle reported that his selection experiments with hooded rats had produced wholly new grades beyond the original limits of variation, a result inconsistent with the predictions of Johannsen, Jennings, and Pearl. In 1914 Castle and his assistant John Phillips reported on the progress of the selection experiments after thirteen consecutive generations of selection.[76] They found that selection had continued to be effective. Moreover, they found that return selection was just as difficult as selecting for extremes because regression occurred toward the new mode established by selection.

At this time Castle believed firmly that he was selecting variations in the Mendelian character itself. Many geneticists disagreed with him. His colleague East, among others, suggested the possibility that he was really selecting modifier genes. The strongest opposition to Castle's view came from the *Drosophila* workers, who found numerous modifiers which were inherited in a Mendelian fashion. By 1915 they had found no less than seven modifiers for the eye color eosin. And when the modifiers were removed the eosin eye color reverted to its original characteristics. They concluded that the Mendelian factors were very stable and that Castle had been selecting for modifiers. Neither Castle nor the *Drosophila* workers denied that selection was effective. The argument was over the basis of the heritable variation upon which selection acted.

In 1916 Castle and Sewall Wright published the results of selection in the hooded rats for three more generations.[77] Se-

76. W. E. Castle and John C. Phillips, *Piebald Rats and Selection,* Carnegie Institution of Washington Publication, no. 195 (Washington, D.C., 1914).

77. W. E. Castle and Sewall Wright, *Studies of Inheritance in Guinea-Pigs and Rats,* ibid., no. 241 (1916).

lection had continued to be effective. This publication also contained the preliminary results of an experiment suggested by Wright to decide the issue of the modifiability of the Mendelian hooded character. Wright's plan was to breed individuals from the plus and minus series with wild, non-hooded rats and to extract the recessive hooded pattern in the F_2 offspring. This process could be repeated by mating the hooded F_2 offspring with wild rats and again extracting the hooded recessive. If Castle had been selecting modifiers, they should be stripped from the hooded trait and the extracted hooded patterns from the plus and minus stocks should again be alike. If the hooded character had been modified, the extracted hooded characters from the plus and minus stocks should retain their differences.

The crosses of the minus races with the wild rats started slowly, but those with the plus race and wild rats were soon successful. After the hooded pattern was extracted three times (six generations) from the wild rats, the mean grade of the hooded character from the plus race dropped only from +3.73 to +3.04 on Castle's scale. Castle again concluded that no modifiers were involved and that the Mendelian character itself had been changed.

The first half of Wright's test experiment seemed to support Castle's position. But the evidence against the modifiability of Mendelian factors was mushrooming in the hands of the *Drosophila* workers. Not only had they found modifiers which behaved as Mendelian factors, but they were able to track some modifiers to the chromosomes on which they were located. Additional evidence was obtained in selection experiments on *Drosophila* by E. C. MacDowell and A. H. Sturtevant.[78] Both found that selection was effective in genetically diverse populations, but their results could only be explained if selection was acting upon modifiers. They showed that a selected inbred strain could not be return selected. This should have been possible if the Mendelian character under selection itself varied. Sturtevant's paper on selection included a forceful criticism of the thirteen most significant cases in which

78. A. H. Sturtevant, *An Analysis of the Effects of Selection,* ibid., no. 264 (1918).

Castle and others claimed that selection had altered a Mendelian factor.

The case against Castle's position was convincing. He tenaciously held to hs view until the results of the crosses of the minus race of hooded rats and wild rats became clear. The results were striking. In three crosses with the wild rats the mean grade of the hooded pattern in the minus series rose from —2.63 to +2.55 On the basis of this result Castle retracted his long held belief in the modifiability of Mendelian factors. In his published retraction he said:

The wild race, which we used in our crosses, evidently had a residual heredity much more like that of our plus-selected than like that of our minus-selected race. When the hooded gene from either race was introduced by repeated crosses into the residual heredity, the result was to produce hooded races of very similar grade, a little lower in grade than the plus selected race, but very much higher in grade than the minus selected race.

It thus becomes clear that the changes which had occurred in the hooded character as a result of selection were *detachable* changes and are probably in nature independently inherited modifying factors.[79]

Although retracting his position on the modifiability of Mendelian factors, Castle believed his selection experiments represented significant progress toward an understanding of the role of selection in evolution. He summarized what he had fought against:

The "Mutation Theory" of de Vries gave us a picture of selection as an agency temporarily effective in producing racial changes, but with those changes gradually vanishing as soon as selection ceased. Johannsen denied within "pure lines" even temporary effectiveness of selection. A strictly logical use of Johannsen's conclusions would have limited their application to such organisms as he studied, self-fertilizing ones completely homozygous for all genetic factors and subject apparently to no new changes in such factors. But the doctrine was straightway extended in the views of most geneticists to selection of every sort and he was treated as a traitor to

79. W. E. Castle, "Piebald Rats and Selection, a Correction," *American Naturalist* 53 (1919): 373.

Mendelism who saw any utility in selection or advocated its use as a means of improving the inherited characters of animals or plants.[80]

But, added Castle, "the situation is wholly different today." This was true, and it was in significant degree a result of the stimulus provided by Castle's selection experiments with hooded rats and his strong support for the selection theory of evolution.

In 1918, because of the developments described above, many geneticists accepted Mendelism and Darwinism as complementary. Some Mendelians, among them Raymond Pearl and R. C. Punnett, were still strongly opposed to the Darwinian selection theory. But their viewpoint was becoming less popular. The recognition in 1918 that Mendelism complemented Darwinism was only a preliminary conclusion. The detailed synthesis of Mendelism and Darwinism by the quantitative investigation of the consequences of Mendelian heredity and of the effects of selection was to be accomplished during the next decade and a half. This new quantitative synthesis of Mendelism and Darwinism was population genetics.

80. Ibid., p. 374.

5

Population Genetics: The Synthesis of Mendelism, Darwinism, and Biometry

WHEN CHARLES DARWIN FIRST ENVISIONED THE PROCESS OF natural selection he believed firmly that a population naturally produced much new heritable variability each generation. He also believed in blending inheritance, which was nature's way of keeping a species true and uniform despite this new variability each generation. Thirty years later he finally designed his provisional hypothesis of pangenesis to account for the production of sufficient variability to make natural selection possible, even though blending inheritance tended to obliterate the variability. But Darwin's theory of pangenesis was never very successful, and the most basic weakness of his concept of evolution was the lack of an adequate theory of the production of the variations upon which natural selection acted. Mendel's theory of heredity was the perfect complement to Darwin's idea of natural selection. Mendelian characters could be very small and were not blended away by crossing. Furthermore, Mendelian recombination provided new variability for selection. When Mendelian heredity was rediscovered, however, for a variety of scientific and personal reasons, it became associated with the mutation theory of evolution rather than with Darwin's idea of continuous evolution. One consequence of this association was that as Mendelism gained attention in the first decade of this century, Darwin's idea of natural selection lost attention. But by 1918, primarily as a result of the analysis of successful selection experiments, many geneticists had realized that Mendelian heredity and Darwinian natural selection were complementary. The study of evolution, however, required more than the general recognition that Mendelism and Darwinism were complementary. It required a careful investigation of the evolutionary consequences of Mendelian heredity. Under a given set of assumptions the mathematical consequences of

Mendelian inheritance could be computed. Thus the study of evolution required a synthesis of Mendelism, Darwinism, and biometry. The basic elements of this synthesis were carried out by R. A. Fisher, J. B S. Haldane, and Sewall Wright.

EXPLORATION OF THE MATHEMATICAL CONSEQUENCES OF MENDELIAN HEREDITY BEFORE 1918

Three developments before 1918 in the exploration of the mathematical consequences of Mendelian heredity influenced the work of Fisher, Haldane, and Wright. The first of these was the (later named) Hardy-Weinberg equilibrium principle, which was of basic importance for population genetics because it guaranteed that variability was preserved in random breeding Mendelian populations. The second was the work on the mathematical consequences of inbreeding carried out primarily in the United States. This work influenced Sewall Wright, who later supplied a powerful analysis of the quantitative aspects of inbreeding. The third development was the analysis of the effects of selection prepared by the mathematician H. T. J. Norton and published in a book by R. C. Punnett. Norton's work stimulated both J. B. S. Haldane and the Russian geneticist Chetverikov to examine further the mathematical consequences of selection under a variety of assumptions about the constitution of the population. These three developments were independent but are here treated together because they partially formed the foundation for the work of Fisher, Haldane, and Wright.

THE HARDY-WEINBERG EQUILIBRIUM PRINCIPLE

The first person to explore the mathematical consequences of Mendelian heredity was Mendel himself. In the paper describing his experiments with peas, in a section entitled "The Subsequent Generations [Bred] from the Hybrids," he calculated the effects of continued self-fertilization on the genetic constitution of a population. Starting with a population formed by the hybridization of races AA and aa, he found as a trial and error generalization that in the nth generation of self-fertilization the ratio of genotypes was $(2^n - 1)AA : 2Aa : (2^n - 1)aa$. Continued self-fertilization clearly led to increasing reversion to the parental types *AA* and *aa,* and to the decrease of heterozygotes.

The next logical step would have been to calculate the effect of Mendelian inheritance for a single locus with two alleles in a random breeding population instead of one which was self-fertilized. Mendel published no analysis of this problem. Had he considered the problem, he probably would have derived the Hardy-Weinberg equilibrium principle. After the rediscovery of Mendelism the first investigator to look at the consequences of Mendelian heredity in a random breeding population was Yule in 1902. He first supposed the existence of two races, one pure for the dominant character *A* and the other pure for the recessive character *a*. Then he asked:

What, exactly, happens if the two races *A* and *a* are left to themselves to inter-cross freely *as if they were one race?* . . . Now when *A*'s and *a*'s are first inter-crossed we get the series of *uniform* hybrids; when these are inter-bred we get the series of three dominant forms (two hybrids, one pure) to one recessive. If all these are again inter-crossed at random the composition remains unaltered. "Dominant" and "recessive" gametes are equally frequent, and consequently conjugation of a "dominant" gamete will take place with a "recessive" as frequently as with another "dominant" gamete.[1]

Yule recognized the stability of the 1*AA*:2*Aa*:1*aa* ratio in a random breeding population. And his method of looking at the population as a gene pool was later widely used in population genetics. He neglected to mention that at this time he believed the 1:2:1 ratio was the *only* stable equilibrium for the gene frequencies, that is, when *A* and *a* were equally numerous in the population.

Yule's paper elicited a response from William Castle in the United States. Castle read the section in which Yule showed the stability of the 1:2:1 ratio and mistakenly thought Yule had claimed that selective elimination of all recessives each generation did not lead toward complete homozygosis in the population. Yule had said no such thing. But with this stimu-

1. Yule, "Mendel's Laws and Their Probable Relations to Intra-racial Heredity," p. 225.

lus, Castle attacked the problem of gene frequencies under random mating. First he agreed with Yule about the stability of the 1:2:1 ratio. Then he showed that the complete elimination of recessives each generation did lead toward homozygosis. This was obvious and Yule would surely have agreed. He also made the conclusion (and here Yule would have disagreed, though this was not clear from his 1902 paper) that "as soon as selection is arrested the race remains stable at the degree of purity then attained."[2] In nonmathematical terms, this was the generalized equilibrium principle for a single locus with two alleles in a random breeding population.

Karl Pearson read the papers of Yule and Castle on the mathematical consequences of Mendelian heredity. In 1904, starting with the assumption that two equally numerous races *AA* and *aa* were randomly bred together, he worked out the equilibrium principle for a single locus with two alleles.[3] He could easily have demonstrated the general result for any initial gene frequencies, instead of $0.5A$ and $0.5a$, but he did not. He did generalize, under the original assumptions, to the case of n loci. He concluded:

> However many couplets we suppose the character under question to depend upon, the off-spring of the hybrids—or the segregating generation—if they breed at random *inter se,* will not segregate further, but continue to reproduce themselves in the same proportions as a stable population.[4]

Pearson was right that a stable equilibrium existed for n loci, but he was wrong that the equilibrium was reached in one generation except for the case $n = 1$.

G. H. Hardy's well-known proof of the equilibrium principle came as the result of a curious encounter. In 1908 R. C. Punnett delivered an address at the Royal Society of Medicine on "Mendelian Heredity in Man."[5] In the subsequent discussion Yule suggested that a dominant allele, once introduced

2. Castle, "The Laws of Heredity of Galton and Mendel," p. 337.
3. Pearson, "On a Generalized Theory of Alternative Inheritance, with Special Reference to Mendel's Laws," pp. 58–60.
4. Ibid., p. 60.
5. For Punnett's account of this address, see his "Early Days of Genetics," p. 9.

into the population, would increase in frequency until reaching stability at 0.5, giving the usual phenotypic ratio 3 dominant:1 recessive thereafter. Punnett knew Yule had to be wrong but did not see how to prove it. He took the problem to his friend Hardy, with whom he formerly played cricket. Hardy immediately derived and published in *Science* on 10 July 1908 the conditions for a stable equilibrium in the case of a single locus with two alleles under random mating.[6] Given that the frequency of genotype $AA = p$, of $Aa = 2q$, and of $aa = r$, he showed that the condition for a stable equilibrium was $q^2 = pr$. The condition for a stable equilibrium was always established by a single generation of random mating; so the distribution would remain unchanged in succeeding generations.

The other independent derivation of the Hardy-Weinberg equilibrium principle was of course given by Wilhelm Weinberg (1862–1937), a German physician who devoted considerable energy to the study of human genetics.[7] He became acquainted with Mendelism in 1905 and decided to see if he could find a character in man which was inherited in a Mendelian fashion. Having conducted numerous twin studies before 1905, he suspected the ability to bear dizygotic twins was a Mendelian trait. In a paper read on 13 January 1908 he derived the general equilibrium principle for a single locus with two alleles.[8] Thus he anticipated Hardy by almost six months with this derivation. He calculated the expected frequencies if the twinning trait were dominant or recessive, and decided his data showed the trait was a Mendelian recessive. The Hardy-Weinberg law, once enunciated, was accepted by all Mendelians. It was an obvious deduction from the mechanics of Mendelian heredity. Because Weinberg's papers were ignored,

6. "Mendelian Proportions in a Mixed Population," *Science,* n. s. 28:49–50.

7. For a short account of the life and work of Weinberg, see Curt Stern, "Wilhelm Weinberg," *Genetics* 44 (1962): 1–5.

8. Wilhelm Weinberg, "Ueber den Nachweis der Vererbung beim Menschen," *Jahreshefte des Vereins für Vaterländische Naturkunde in Württemburg* 64 (1908): 368–82. English translation in Samuel H. Boyer, *Papers on Human Genetics* (Englewood Cliffs, N.J.: Prentice-Hall, 1963), pp. 4–15.

the law was for many years known to geneticists as "Hardy's Law."

Weinberg did not stop with the equilibrium principle which now carries his name. In 1904 Pearson had concluded that Mendelism was incompatible with observed correlations in human populations. Weinberg attempted to demonstrate that Pearson was wrong. In 1909 and 1910 he published three other papers on the mathematical consequences of Mendelian heredity with special reference to human genetics.[9] He derived the correlations to be expected between close relatives on the basis of Mendelian inheritance for several cases and made quantitative provision for environmental influences. Fisher, unaware of Weinberg's work, was to make similar calculations in a paper published in 1918. Weinberg extended the equilibrium law in random breeding populations to cover the case of a single locus with multiple alleles. He also extended the law to more than one locus and discovered that equilibrium was not attained in a single generation as Pearson had concluded.

Unfortunately, as mentioned above, Weinberg's papers were ignored by geneticists. Few Mendelians knew enough mathematics to appreciate what he was doing. The biometricians were offended by Weinberg's attacks. Pearson wrote in 1909 about Weinberg's "curiously ignorant account of the biometric treatment of heredity" and said "it hardly seems needful to reply to criticisms of this character."[10] Some geneticists did not read German. Others inexplicably ignored Weinberg's work. Raymond Pearl, a Mendelian with the necessary mathematical background, had a paper published adjacent to one of Weinberg's in 1909 in *Zeitschrift für Induktive Abstammungs- und Vererbungslehre;* yet he made no reference to Weinberg's work. By the time Weinberg's efforts were appreciated, popu-

9. Wilhelm Weinberg, "Ueber Vererbungsgesetze beim Menschen. 1. Allgemeiner Teil," *Zeitschrift für Induktive Abstammungs- und Vererbungslehre,* 1 (1909): 377–92, 440–60; "Ueber Vererbungsgesetze beim Menschen. 2. Spezieller Teil," ibid., 2 (1909): 276–330; "Weitere Beiträge zur Theorie der Vererbung," *Archiv für Rassen- und Gesellschafts- Biologie* 7 (1910): 35–49, 169–73.

10. Pearson, "Darwinism, Biometry, and Some Recent Biology," p. 381.

lation genetics had already surpassed them in sophistication. Thus Weinberg had little influence upon the rise of population genetics, though he was a pioneer in the field.

THE QUANTITATIVE ANALYSIS OF INBREEDING

In the United States between 1912 and 1918 the study of the mathematical consequences of Mendelian heredity centered primarily upon the problems of inbreeding—long a subject of controversy among biologists and breeders. Some claimed inbreeding was deleterious while others claimed it was necessary to preserve desired traits in new varieties or breeds. Many geneticists realized that if a population were of known genetic composition and operated under Mendelian inheritance, one should be able to calculate the genetic composition of the population in future generations if fluctuations due to sampling were ignored. In this way the genetic consequences of inbreeding could be analyzed.

Between 1912 and 1916 H. S. Jennings and Raymond Pearl each published a series of papers on the quantitative analysis of the consequences of systems of inbreeding under Mendelian inheritance.[11] These papers attracted considerable attention and stimulated others by Wentworth and Remick [12] and Robbins.[13] The problem with all this research was the method used. The researchers assumed a certain distribution of genotypes in one generation, then laboriously calculated the distribution of genotypes in the next generation. Following this they attempted to derive by trial and error a formula for the distribution of genotypes in future generations. The method

11. H. S. Jennings, "Production of Pure Homozygotic Organisms from Heterozygotes by Self-Fertilization," *American Naturalist* 46 (1912): 487–91; "Formulae for the Results of Inbreeding," ibid., 47 (1914): 693–96; "The Numerical Results of Diverse Systems of Breeding," *Genetics* 1 (1916): 53–89. Raymond Pearl, *Modes of Research in Genetics* (New York: Macmillan, 1915), pp. 101–56. In this book Pearl summarized his research on inbreeding carried out in the years 1913–15.

12. E. N. Wentworth and B. L. Remick, "Some Breeding Properties of the Generalized Mendelian Population," *Genetics* 1 (1916): 608–16.

13. Rainard B. Robbins, "Some Applications of Mathematics to Breeding Problems," *Genetics* 2 (1917): 489–504; 3 (1918), 73–92, 375–89; "Random Mating with the Exception of Sister by Brother Mating," ibid., pp. 390–96.

worked well for simple systems of mating but became progressively cumbersome with more complex systems. Thus by 1918 the mathematical study of systems of inbreeding had reached an impasse. When the problems of inbreeding attracted the serious attention of Sewall Wright in about 1919, he abandoned the method which had led to the impasse and developed a more sophisticated method for the mathematical analysis of inbreeding. Wright's work on inbreeding was important because it deeply influenced his later mathematical analysis of evolution in nature.

THE MATHEMATICAL ANALYSIS OF SELECTION

Jennings had considered some simple cases of selection in his papers of 1916 and 1917. But a more influential account of the effects of selection in a Mendelian population had already appeared in England in 1915 in a book entitled *Mimicry in Butterflies* [14] by R. C. Punnett. His argument was that mimicry was an evolutionary phenomenon which must occur by distinct leaps: "How can we suppose that a slight variation in the direction of the model on the part of the [would-be mimic] would be of any value to it?" [15] Supposing the distinct leap were caused by a Mendelian factor, Punnett wanted to know how fast selection would cause the factor to spread through or be eliminated from the population. Accordingly, he requested the mathematician H. T. J. Norton of Trinity College, Cambridge, to prepare a table for him showing the effects of selection of various intensities acting upon a Mendelian factor in a random breeding population. Norton's table, reproduced here (see table 2), showed the number of generations required, at selection intensities of 0.50, 0.25, 0.10, and 0.01, to change the gene frequency from various intervals between 0.999 and 0.000. Punnett noted that a recessive trait with a selective disadvantage of 0.10 would be reduced in frequency from 0.44 to 0.028 in only 70 generations and that with a selective disadvantage of 0.01 the same reduction of gene frequency would require about 700 generations.

Punnett was impressed by the effectiveness of selection:

14. (Cambridge: Cambridge University Press, 1915).
15. Ibid., p. 62.

TABLE 2

Selection Table Prepared for R. C. Punnett by H. T. J. Norton

| Percentage of total population formed by old variety | Percentage of total population formed by the hybrids | Percentage of total population formed by the new variety | Number of generations taken to pass from one position to another as indicated in the percentages of different individuals in left-hand column | | | | | | | |
| | | | A. Where the new variety is dominant | | | | B. Where the new variety is recessive | | | |
			$\frac{100}{50}$	$\frac{100}{75}$	$\frac{100}{90}$	$\frac{100}{99}$	$\frac{100}{50}$	$\frac{100}{75}$	$\frac{100}{90}$	$\frac{100}{99}$
99·9	·09	·000	4	10	28	300	1920	5740	17,200	189,092
98·0	1·96	·008	2	5	15	165	85	250	744	8,160
90·7	9·0	·03	2	4	14	153	18	51	149	1,615
69·0	27·7	2·8	2	4	12	121	5	13	36	389
44·4	44·4	11·1	2	4	12	119	2	6	16	169
25·	50·	25·	4	8	18	171	2	4	11	118
11·1	44·4	44·4	10	17	40	393	2	4	11	120
2·8	27·7	69·0	36	68	166	1,632	2	6	14	152
·03	9·0	90·7	170	333	827	8,243	2	6	16	165
·008	1·96	98·0	3840	7653	19,111	191,002	4	10	28	299
·000	·09	99·9								

NOTE: Reprinted by permission of the publisher from R. C. Punnett, *Mimicry in Butterflies* (Cambridge: Cambridge University Press, 1915), p. 155.

Evolution, in so far as it consists of the supplanting of one form by another, may be a very much more rapid process than has hitherto been suspected, for natural selection, if appreciable, must be held to operate with extraordinary swiftness where it is given established variations with which to work.[16]

Norton's table also showed that selection was ineffective when acting against a rare recessive. In 1917 Punnett used this idea to discredit the claim of eugenicists that deleterious recessives could be eliminated from human populations in a few generations.[17]

Norton's table was the perfect complement to Morgan's theory of evolution by single gene replacement and it had a widespread influence. J. B. S. Haldane was stimulated by the table as was the Russian geneticist Chetverikov. Curiously, the implications of Norton's table were to undermine the discontinuous theory of evolution expounded by Punnett. The table showed clearly that small selection intensities acting for surprisingly small numbers of generations could greatly change gene frequencies in a population.

In 1917 Howard C. Warren, a Princeton psychologist, presented a short mathematical argument that Mendelism and Darwinian selection were compatible.[18] Warren treated a random breeding population for two special cases: when the dominant phenotype was twice as viable as the recessive, and when the recessive was twice as viable as the dominant. He found mathematically that selection should be effective, though the elimination was more rapid when the recessive was more viable. Warren's conclusion was that Mendelism and selection together formed a likely basis for evolutionary change. When Haldane published his first paper on mathematical selection theory in 1924 he cited Norton's table and Warren's paper as the only previous publications on the topic.

By 1918 the setting was complete for a synthesis of Mende-

16. Ibid., p. 96.
17. R. C. Punnett, "Eliminating Feeblemindedness," *Journal of Heredity* 8 (1917): 464–65.
18. Howard C. Warren, "Numerical Effects of Natural Selection Acting upon Mendelian Characters," *Genetics* 2 (1917): 305–12.

lian heredity, Darwinian selection, and biometrical methods. Mendelism was widely accepted. For explaining the success of selection acting upon small differences, Mendelism and Darwinian selection were recognized as complementary. All that remained was the quantitative synthesis with biometrical methods, some of which had already been applied to an analysis of the consequences of Mendelian heredity.

RONALD ALYMER FISHER

R. A. Fisher (1890–1962) exhibited a special aptitude for mathematics and astronomy at an early age. His mother read him elementary astronomy before he was six, and several years later he was attending lectures on astronomy by Sir Robert Ball. In preparatory school teachers recognized his abilities and encouraged his mathematical investigations. While attending the Harrow School, Fisher was instructed by W. N. Roseveare, often in the evenings. Since Fisher suffered from extreme myopia and was prohibited from working by electric light, Roseveare usually taught him without visual aids—a practice that heightened Fisher's ability to visualize and solve mathematical problems in his head. Later, some mathematical statisticians and geneticists were to complain that Fisher's proofs contained intuitive leaps which were not obvious.

In 1909 Fisher entered Gonville and Caius College, Cambridge, on a scholarship. He excelled in mathematics. His tutor, the astronomer F. J. M. Stratton, encouraged him while still an undergraduate to publish a paper entitled "On an Absolute Criterion for Fitting Frequency Curves."[19] Stratton then urged Fisher to send a copy to a friend, the mathematician W. S. Gosset, who published papers under the name "Student." Gosset described his reaction to Fisher's paper in a letter to Karl Pearson on 12 September 1912:

Stratton, the tutor, made him [Fisher] send me [the paper] and with some exertion I mastered it, spotted the fallacy (as I believe) and wrote him a letter showing, I hope, an intelligent interest in the matter and incidentally making a blunder. To this he replied with two foolscap pages covered with math-

19. *Messenger of Mathematics* 41 (1912): 155–60.

ematics of the deepest dye in which he proved [what he had previously claimed] and of course exposed my mistake. I couldn't understand his stuff and wrote and said I was going to study it when I had time.[20]

Despite Gosset's less than encouraging reply, Fisher wished to continue corresponding with him because they were interested in similar mathematical investigations. In 1908 Gosset had published without proof the exact solution for the test of the significance of the mean of a small sample of normally distributed material. Fisher took up the problem and by early September 1912 he had derived a rigorous proof of Gosset's solution. He sent his calculations to Gosset, who again could not understand them. Gosset sent Fisher's proof on to Pearson with the comment, "it seems to me that if it's all right perhaps you might like to put the proof in a note [in *Biometrika*]. It's so nice and mathematical that it might appeal to some people."[21] Clearly by 1912, at age twenty-two, Fisher was already an accomplished mathematician.

While at Cambridge, Fisher's interests were not confined to pure mathematics. He discovered Karl Pearson's series "Mathematical Contributions to the Theory of Evolution" and became interested in genetics and evolution. The Mendelians were influential at Cambridge and Fisher, unlike Pearson, soon became convinced that Mendelism was the prevailing mechanism of heredity. He also studied astronomy at Cambridge. After graduating in 1912 he stayed for another year with a studentship in physics, studying under James Jeans and his former tutor Stratton. In the years following his departure from Cambridge in 1913, Fisher's primary interests were in the fields of mathematical statistics and evolutionary theory.

For several years before 1914, Pearson and his colleagues had been concerned with the reliability of a correlation coefficient derived from a sample of a much larger population. In September 1914, Fisher solved the problem by deriving the exact

20. W. S. Gosset to Karl Pearson, 12 September 1912. Reprinted in E. S. Pearson, "Studies in the History of Probability and Statistics. 20. Some Early Correspondence between W. S. Gosset, R. A. Fisher, and Karl Pearson, with Notes and Comments," *Biometrika* 55 (1968): 446.
21. Ibid.

distribution of the values of the correlation coefficient in samples from an indefinitely large population. He sent the derivation to Pearson who wrote back on 26 September 1914 to say "I congratulate you very heartily on getting out the exact distribution form of r. . . . If the analysis is correct which seems highly probable, I should be delighted to publish the paper in *Biometrika*." [22]

Fisher's derivation of the exact distribution of the correlation coefficient was of little value in practical usage without the computation of tables of the distribution of r for various sample sizes. Thus even before the appearance of Fisher's paper in *Biometrika* in May 1915 both Fisher and Pearson (with the resources of his statistical laboratory) were working on tables for the distribution of r. Unfortunately the war prevented easy communication between Fisher and the workers at Pearson's statistical laboratory. The breakdown in communications allowed Pearson's group to proceed without understanding some parts of Fisher's 1915 paper and caused Fisher some unnecessary work. Relations between the two men began to be strained. On 15 May 1916 Fisher wrote Pearson that "I could probably have worked more profitably, if I had been in closer touch with the Laboratory, although such collaboration is never easy." [23]

The May 1916 issue of *Biometrika* contained a paper by Kirstine Smith on the "best" values of the constants in frequency distributions. Fisher disagreed with her conclusions and immediately wrote a note of rebuttal which he submitted to Pearson for publication in *Biometrika*. In the same letter with the note he informed Pearson: "I have recently completed an article on Mendelism and Biometry which will probably be of interest to you. I find on analysis that the human data is as far as it goes, not inconsistent with Mendelism. But the argument is rather complex." [24] Pearson replied on 26 June that he disagreed with Fisher's criticism of Miss Smith and declined to publish the note. [25] From his experience with the Mende-

22. Pearson to Fisher, ibid., p. 448.
23. Fisher to Pearson, ibid., p. 451.
24. Ibid., p. 454.
25. Pearson to Fisher, ibid., p. 455.

lians Pearson understandably wanted to avoid controversies. In his reply he did not mention Fisher's paper on Mendelism and biometry, but he cannot have agreed with Fisher's thesis. When Fisher's paper was finally published in October 1918, he immediately sent Pearson a copy and received this reply: "Many thanks for your memoir which I hope to find time for. I am afraid I am not a believer in cumulative Mendelian factors as being the solution of the heredity puzzle."[26] Thus Pearson flatly rejected Fisher's thesis even before reading the paper. After 1918 the complications arising from the disagreements between Fisher and Pearson strained relations between them beyond repair.

Fisher's approach to Mendelism and biometry was that advocated by Yule more than ten years earlier. He wanted to synthesize Darwinism, Mendelism, and biometry. Probably by 1912 Darwin's *Origin of Species* and Pearson's papers had convinced Fisher that natural selection was the primary agent of evolutionary change and that it operated upon apparently continuous variations. In this sense Fisher was a firm Darwinian. But he disagreed with Pearson's and Darwin's analysis of continuous variation. Pearson claimed, and Darwin would probably have agreed, that the continuous variations in a pure line were heritable and that continued selection in a pure line should be effective. Because blending inheritance eliminated much of the heritable variation each generation Darwin believed that many new continuous variations must be heritable. Otherwise the supply of heritable variation in a population would be drastically depleted within a few generations. Probably as a result of the influence of the Cambridge Mendelians, Fisher, unlike Pearson, believed in Mendelian inheritance and the continuity of the germ plasm. He quickly realized, along with Bateson, de Vries, and others, that individual factors were not blended away by crossing. He saw that the mathematical consequences of Mendelian heredity in general preserved Mendelian factors, and thus heritable variability, in the population. He knew that selection in a pure line with genetically identical individuals must be ineffective. It followed that much

26. Pearson to Fisher, 21 October 1918, ibid., p. 456.

of the continuous variation in a heterogeneous population where selection was effective must be capable of explanation in Mendelian terms. Fisher set out to demonstrate what Yule had suggested and Weinberg proved in some detail: that Mendelism could account for observed correlations between relatives, despite Pearson's belief to the contrary. He apparently was unaware of Weinberg's work.

Fisher completed his paper on Mendelism and biometry by June 1916 and submitted the paper to the Royal Society of London for publication. The referees suggested it be withdrawn. He subsequently submitted the paper to the Royal Society of Edinburgh, which with his financial assistance published it on 1 October 1918 under the title "The Correlation between Relatives on the Supposition of Mendelian Inheritance."[27]

Fisher's express purpose in the paper was to interpret the well-established results of biometry in terms of Mendelian inheritance by ascertaining the biometrical properties of a Mendelian population. In particular, he wanted to show that Pearson was mistaken in concluding that the correlations between relatives in man contradicted the Mendelian scheme of inheritance. He began by defining a measure of the variability of a character in a population. Often the standard deviation σ was utilized for this purpose. But Fisher noted that

when there are two independent causes of variability capable of producing in an otherwise uniform population distributions with standard deviations σ_1 and σ_2, it is found that the distribution, when both causes act together, has a standard deviation $\sigma_1^2 + \sigma_2^2$. It is therefore desirable in analysing the causes of variability to deal with the square of the standard deviation as the measure of variability. We shall term this quantity the Variance of the normal population to which it refers, and we may now ascribe to the constituent causes fractions or percentages of the total variance which they together produce.[28]

The paper was devoted to an analysis of the constituent parts of the total variance in a Mendelian population.

27. *Transactions of the Royal Society of Edinburgh* 52 (1918): 399–433.
28. Ibid., p. 399.

He pointed out that the observed correlation between relatives was an unreliable direct measure of the percentage of the total variance contributed by ancestors:

> For stature the coefficient of correlation between brothers is about .54, which we may interpret by saying that 54 per cent of their variance is accounted for by ancestry alone, and that 46 per cent must have some other explanation.
>
> It is not sufficient to ascribe this last residue to the effects of environment. Numerous investigations by Galton and Pearson have shown that all measurable environment has much less effect on such measurements as stature. Further, the facts collected by Galton respecting identical twins show that in this case, where the essential nature is the same, the variance is far less. The simplest hypothesis, and the one which we shall examine, is that such features as stature are determined by a large number of Mendelian factors, and that the large variance among children of the same parents is due to the segregation of those factors in respect to which the parents are heterozygous. Upon this hypothesis we will attempt to determine how much more of the variance, in different measurable features, beyond that which is indicated by the fraternal correlation, is due to innate and heritable factors.[29]

Fisher knew that some influences tended to obscure the actual genetic similarity between relatives. Dominance could cause different somatic effects with identical genetic changes. Genic interaction, or epistasis, could also cause this. Fisher termed these genetic processes "nonadditive." Thus he divided the total genetic contribution from one generation to another into an additive part and a nonadditive residue.

From the pure line work it was obvious that environmental influences also tended to obscure the actual genetic similarity between relatives. In 1906 Yule had suggested that the effects of incomplete dominance and the environment, when taken into account, would show that Mendelian heredity and observed correlations between human relatives were compatible. Pearson had claimed they were incompatible. Yule also made the comment that "so far as the coefficients of correlation alone are concerned, it is . . . impossible to distinguish between the effects of the heterozygote giving rise to forms that are not

29. Ibid., p. 400.

strictly intermediate, and the effect of the environment in causing somatic variations which are not heritable." [30] In other words, it was impossible to distinguish the effects of environment from the effects of dominance in the correlations between relatives. Fisher showed that Yule was wrong; these effects could be distinguished.

It was well known by the biometricians that fraternal correlation usually exceeded parental. Noting that the variance in a sibship, apart from environmental effects, depended only upon the number of factors in which the parents are heterozygous, Fisher calculated the fraternal correlation as compared to parental correlation. The calculation showed that

the effect of dominance is to reduce the fraternal correlation to only half the extent to which the parental correlation is reduced. This allows us to distinguish, as far as the accuracy of the existing figures allows, between the random external effects of environment and those of dominance.[31]

By analyzing the extent to which fraternal correlation exceeded parental, he was able to distinguish the contributions of dominance and environment to the total variance.

In addition to accounting for the effects of dominance, Fisher examined the statistical consequences of genic interaction, assortative mating, multiple alleles, and linkage upon the correlations between relatives. He believed the effects of genic interaction and linkage were negligible in a large population. He extended his analysis to the correlations between uncles and cousins and other relatives. Then, using the data of Pearson and Lee on man,[32] from which Pearson had concluded the inadequacy of Mendelian inheritance, Fisher demonstrated that "the hypothesis of cumulative Mendelian factors seems to fit the facts very accurately." [33] One important conclusion he made from the application of his theory to the data of Pearson and Lee was that "it is very unlikely that so much

30. Yule, "On the Theory of Inheritance," p. 142.
31. Fisher, "The Correlations between Relatives," p. 406.
32. Karl Pearson and Alice Lee, "On the Laws of Inheritance in Man. 1. Inheritance of Physical Characters," *Biometrika* 2 (1903): 357–462.
33. Fisher, "The Correlations between Relatives," p. 433.

as 5 per cent of the total variance is due to causes not heritable." [34] Fisher concluded that many continuously varying characters such as human stature were primarily determined by many Mendelian factors not environmental influences.

Fisher's 1918 paper was well received by the few geneticists who could understand his mathematics. Encouraged, he next attempted the task from which William Bateson had shrunk: to quantitatively examine the evolutionary consequences of Mendelian heredity. In 1922 he published a substantial paper on this topic.[35] He discussed the interaction of selection, dominance, mutation, random extinction of genes, and assortative mating. The germinal ideas of many of his later researches into evolution were contained in this paper. First he treated the problem of equilibrium under selection. For a single locus with two alleles, he showed that if selection favored one homozygote the other allele would be eliminated. He then stated the possibility of a balanced polymorphism and its consequences:

If, on the other hand, the selection favors the heterozygote, there is a condition of stable equilibrium, and the factor will continue in the stock. Such factors should therefore be commonly found, and may explain instances of heterozygote vigor, and to some extent the deleterious effects sometimes brought about by inbreeding.[36]

Next he considered the problem of the survival of individual genes. He found that individually a gene had a very small chance of surviving. The survival of a rare gene depended upon chance rather than selection. A mutation would be more likely to become fixed at low frequencies in a large instead of a small population simply because the mutation would more often survive in a large population. "Thus a numerous species, with the same frequency of mutation, will maintain a higher variability than will a less numerous species: in connection with this fact we cannot fail to remember the dictum of Charles Darwin that 'wide ranging, much diffused and com-

34. Ibid., p. 524.
35. R. A. Fisher, "On the Dominance Ratio," *Proceedings of the Royal Society of Edinburgh* 42 (1922): 321–41.
36. Ibid., p. 324.

mon species vary most.'"[37] A consequence of this point of view was that a smaller mutation rate could balance the effects of adverse selection in a large population more easily than in a small population. This idea was a fundamental tenet of Fisher's view of evolution.

In 1921 A. L. and A. C. Hagedoorn had published the theory that the random survival of genes in populations was more important than preferential survival as a result of selection.[38] Attacking this idea with vigor, Fisher demonstrated that even with the absence of new mutations in a population of moderate size (about 10,000 individuals) the rate of gene extinction was exceedingly small. He therefore rejected the importance of the chance elimination of genes as compared with the elimination by selection.

If the heterozygote were intermediate between the homozygotes at a locus, Fisher showed that selection could quickly eliminate one allele. But in the case of complete dominance, selection was ineffective in removing deleterious recessives present at low frequencies. Thus under the protection of dominance there was an accumulation of rare recessives in the population. This effect was heightened in large populations because a low frequency of mutation could sustain the presence of an allele.

In the 1918 paper Fisher defined the quantity a^2 as the contribution which a single locus makes to the total variance.[39] He now concluded on the basis of his calculations that one

effect of selection is to remove preferentially those factors for which a is high, and to leave a predominating number in which a is low. In any factor a may be low for one of two reasons: (1) the effect of the factor on development may be very slight, or (2) the factor may effect changes of little adaptive importance. It is therefore to be expected that the large and easily recognised factors in natural organisms will be of little adaptive importance, and that the factors affecting important adaptations will be individually of very slight effect. We should thus expect that variation in organs of adaptive

37. Ibid.
38. A. L. and A. C. Hagedoorn, *The Relative Value of the Processes Causing Evolution* (The Hague: Martinus Nyhoff, 1921).
39. Fisher, "The Correlations between Relatives," p. 402.

importance should be due to numerous factors, which individually are difficult to detect.[40]

Fisher's basic ideas concerning the process of evolution were expressed in this paper. He believed, in accordance with his biometrical training, that evolution was primarily concerned with large populations where variability, because of storage of genes, was high. In such populations the deterministic results of selection acting upon single gene effects reigned supreme. Natural selection was slow but sure. Fisher even went so far as to compare the rules governing evolutionary change to the general laws of the behavior of gases. The investigation of natural selection

may be compared to the analytic treatment of the Theory of Gases, in which it is possible to make the most varied assumptions as to the accidental circumstances, and even the essential nature of the individual molecules, and yet to develop the general laws as to the behavior of gases, leaving but a few fundamental constants to be determined by experiment.[41]

Among the negligible "assumptions as to the accidental circumstances" in evolutionary theory were the effects of genic interaction and random genetic drift. Sewall Wright was to disagree with Fisher's judgment in these cases. Fisher's theory of evolution, like Morgan's, was based upon single gene replacements. Fisher, however, emphasized the smallness of the variations and the slowness of natural selection far more than Morgan, who thought in terms of sizable mutations and relatively rapid selection.

Between 1922 and 1929 Fisher published a series of papers amplifying or experimentally verifying aspects of the evolutionary view presented in his 1922 paper. In 1926 he and E. B. Ford published a study of thirty-five species of British moths.[42] They found, in accordance with Fisher's 1922 prediction, that in one locality the abundant species exhibited much more variability than the rare species with respect to a continuously vari-

40. Fisher, "On the Dominance Ratio," p. 334.
41. Ibid., pp. 321–22.
42. R. A. Fisher and E. B. Ford, "Variability of Species," *Nature* 118 (1926): 515–16.

able character. Fisher considered this data an excellent verification of his theory.

Fisher's theory of evolution harmonized with Darwinian evolution rather than discontinuous evolution. In 1927 one last bastion of the adherents of discontinuous evolution was mimicry theory. In England R. C. Punnett was the champion of discontinuous evolution in mimicry. His 1915 book *Mimicry in Butterflies* was widely read by entymologists. Punnett's theory of mimicry was distasteful to Fisher because it exemplified, he felt, the wrong application of Mendelism to discontinuous evolution. On 5 October 1927 Fisher read a paper before the Entymological Society of London in which he attacked discontinuous evolution in mimicry. In the introduction he stated the general problem:

> It is now becoming increasingly widely understood that the bearing of genetical discoveries, and in particular of the Mendelian scheme of inheritance, upon evolutionary theory is quite other than that which the pioneers of Mendelism originally took it to be. These were already, at the time of the rediscovery of Mendel's work, in the full current of that movement of evolutionary thought, which in the nineties of the last century, had set in favour of discontinuous origin for specific forms. It was natural enough, therefore, that the discontinuous elements in Mendelism should, without sufficiently critical scrutiny, have been interpreted as affording decisive evidence in favour of this view. . . . It should be borne in mind that the reinterpretation of the significance of Mendelism in cases of mimicry is but part of a more general recovery of genetical opinion from positions adopted at a somewhat immature stage of the development of that science.[43]

In the body of the article Fisher first carefully examined the foundation of mimicry theory, especially the ideas of Henry Bates and Fritz Müller and the revisions suggested by G.A.K. Marshall, and then Punnett's theory. Punnett's proof of discontinuity seemed convincing. He had found (in 1915) that the differences among three forms of the butterfly *Papilio polytes* in Ceylon were caused by an apparently stable polymorph-

43. R. A. Fisher, "On Some Objections to Mimicry Theory; Statistical and Genetic," *Transactions of the Entymological Society of London* 75 (1927): 269.

ism involving two Mendelian factors, one of which was necessary for the manifestation of the second. The differences among the three forms were clearly discontinuous; so, in Punnett's view, the forms must have originated by discontinuous leaps.

Punnett assumed without mention that the phenotypic manifestation of the Mendelian factors involved had always been the same. Fisher challenged this assumption. Citing Castle's experiments in which the expression of the factor for the hooded pattern in rats had been significantly changed by the accumulation of modifiers, he suggested that a similar accumulation of modifiers could have changed the expression of the factors involved in the polymorphism in *Papilio polytes*. Generally, in a population where a stable polymorphism existed, if "selection favours different modifications of the two genotypes, it may become adaptively dimorphic by the cumulative selection of modifying factors, without alteration of the single-factor mechanism by which the polymorphism is maintained." [44] Thus Fisher provided a theory of mimicry based upon the deterministic effect of selection acting upon small modifiers. In so doing he helped bring down the last major stronghold of some Mendelians who advocated the discontinuous, anti-Darwinian theory of evolution.

The phenomenon of dominance had long troubled geneticists. Not only was the physiological mechanism of dominance obscure, but also the mechanism of the evolution of dominance was unclear. Fisher became interested in these problems and in 1928 he published a theory of the evolution of dominance.[45] The theory was based upon his conviction that the populations important in evolution were large and that very small selection pressures exerted over long periods of time were crucial in species change. He argued that most mutations tended to be deleterious and to occur at a finite rate. Thus in a large population over a long period of time a mutant

44. Ibid., p. 277.
45. R. A. Fisher, "The Possible Modification of the Response of the Wild Type to Recurrent Mutations," *American Naturalist* 62 (1928): 115–26.

allele was likely to become fixd at low frequencies in the population. Selection against the mutant allele would be balanced by recurrent mutation. Initially, the heterozygote between the new mutant allele and the wild type allele would be intermediate between the homozygous types, in expression and fitness. Since the heterozygotes would be much more frequent in the population than the mutant homozygotes, selection would tend to preserve those heterozygotes which, because of modifying factors, more closely resembled the homozygous wild type. By selection of modifier factors the heterozygote would eventually become phenotypically indistinguishable from the wild type, thus accomplishing dominance. The selection pressures acting upon the modifiers were small, of the order of mutation rates, but Fisher believed such selection pressures were effective given enough time. Both Wright and Haldane disagreed with Fisher's theory of the evolution of dominance, and this disagreement illuminates the way all three differed in their ideas of evolution.

By 1929 Fisher believed he had worked out a relatively complete theory of evolution, synthesizing Mendelian inheritance and Darwinian selection. He put forth his theory in *The Genetical Theory of Natural Selection*,[46] published in 1930. He later stated that one reason he wrote the book was "to demonstrate how little basis there was for the opinion . . . that the discovery of Mendel's laws of inheritance was unfavorable, or even fatal, to the theory of natural selection."[47] In the first chapter Fisher demonstrated mathematically that the consequence of Darwin's assumption of blending inheritance was that "the heritable variance is approximately halved in every generation."[48] Thus Darwin's theory required the appearance of an enormous amount of new variation each generation. Fisher showed that Mendelian inheritance offered a solution to this problem in Darwin's theory because it conserved the variance in the population. In this chapter he also challenged

46. (Oxford: Clarendon Press, 1930).
47. R. A. Fisher, "Retrospect of the Criticisms of the Theory of Natural Selection," *Evolution as a Process,* ed. Julian Huxley, A. C. Hardy, and E. B. Ford (New York: Collier Books, 1963), p. 104.
48. Fisher, *The Genetical Theory of Natural Selection,* p. 5.

the argument used by de Vries, Bateson, Punnett, and many others that small heritable variations could have no selective advantage. Fisher argued that

if a change of 1mm. has selection value, a change of 0.1 mm. will usually have a selection value approximately one-tenth as great, and the change cannot be ignored because we deem it inappreciable. The rate at which a mutation increases in numbers at the expense of its allelomorph will indeed depend upon the selective advantage it confers, but the rate at which a species responds to selection in favour of any increase or decrease of parts depends on the total heritable variance available, and not on whether this is supplied by large or small mutations.[49]

The second chapter introduced Fisher's "fundamental theorem of natural selection." The basic idea was that the effectiveness of selection depended upon the total heritable variance available in the population at that time. If fitness were measured by the ability of a gene to survive and be represented in future generations, then natural selection tended to increase the total fitness of the population. The rate of increase in fitness depended upon the amount of genetic variance in fitness available. Fisher derived his fundamental theorem mathematically, then stated it in words as: *"The rate of increase in fitness of any organism at any time is equal to its genetic variance in fitness at that time."* [50] He compared this general formulation of the action of natural selection not only with the gas laws but also with the second law of thermodynamics. A population always increased in fitness according to Fisher's fundamental theorem, but the process could be continued indefinitely because of the deterioration of the environment, such as increase in mutagens, changes in the physical environment, or increase of the population causing more intense intraspecific competition.

The next five chapters of Fisher's book were devoted to restatement and expansion of views he had expressed in his earlier papers. In the concluding five chapters he extended his

49. Ibid., pp. 15–16.
50. Ibid., p. 35.

genetical ideas to human populations. In 1930 the *Genetical Theory of Natural Selection* represented the most substantial contribution to the synthesis of Mendelism, Darwinism, and biometry yet published.

SEWALL WRIGHT

Sewall Wright approached the problem of evolution with a background different from Fisher's.[51] His interest in evolution was spurred by reading Vernon L. Kellogg's *Darwinism Today*[52] when he was an undergraduate at Lombard College in Galesburg, Illinois. Unlike Fisher, his formal mathematical training was sparse, extending only so far as elementary differential and integral calculus, although he later taught himself a great deal of mathematics as the necessity arose in his quantitative work on evolution. He went to the University of Illinois in 1911 for graduate study in biology. During that year William Castle, then at the Bussey Institution of Harvard, came to give a talk on genetics. Wright was fascinated with his ideas. Since there was no opportunity at this time for him to study genetics at Illinois, Castle encouraged him to transfer to the Bussey Institution. Wright did so the following year and remained there for the years 1912–15.

His work in Castle's laboratory centered upon physiological genetics. He conducted an extensive study of the inheritance of color and other coat characters in guinea pigs and found that the inheritance of color was controlled by an interaction system of genes.[53] This result was important in the development of Wright's later thought on selection and evolution. He was convinced that interaction systems of genes were important in the life of organisms. Thus what was important in evolution was the fate of interaction systems, not just single genes. From the beginning of his work in genetics Wright was

51. I am indebted to Sewall Wright for allowing me two lengthy interviews. His comments elucidated not only his own development but also the work of Fisher, Haldane, and many others.

52. (New York: Henry Holt, 1907).

53. Sewall Wright, *An Intensive Study of the Inheritance of Color and of Other Coat Characters in Guinea-Pigs, with Especial Reference to Graded Variations,* Carnegie Institution of Washington Publication, no. 241 (Washington, D.C., 1916), pp. 59–60.

more concerned with interaction effects than Fisher or Haldane.

While at the Bussey Institution, Wright also helped Castle with the selection experiment on hooded rats. He was fully convinced of the effectiveness of selection in altering a character beyond its former limits of variation. The crucial experiment which demonstrated conclusively that Castle had been selecting modifying factors rather than a single varying factor was suggested by Wright.

Wright had only one brush with population genetics while at the Bussey Institution. In 1913 Raymond Pearl published his first paper on the mathematical consequences of inbreeding in a Mendelian population. He claimed that there was absolutely no "automatic increase in the proportion of homozygotes necessarily following any other sort of inbreeding except self-fertilization" [54] and tried to demonstrate that continued brother-sister mating caused no increase in the proportion of homozygotes. H. D. Fish, a student in the same laboratory as Wright, saw (along with others) that Pearl must be wrong. But Fish knew little mathematics and covered a huge number of scratch pages with calculations without disproving Pearl's assertions. Finally he so cluttered up the laboratory with his pages of calculations that Wright felt compelled to help him. Wright quickly worked out a formula for calculating the composition of the population at any generation, which Fish used to show that continued brother-sister mating indeed caused the population to become progressively more homozygous. Much to Pearl's chagrin, Fish then published his results.[55] Wright participated in this little adventure simply for diversion. At the time he had no inkling that six years later he would devise a powerful general method for the analysis of the consequences of inbreeding in Mendelian populations.

In 1915 Wright left the Bussey Institution for the Animal Husbandry Divison of the United States Department of Agri-

54. Raymond Pearl, "A Contribution towards an Analysis of the Problem of Inbreeding," *American Naturalist* 47 (1913): 606.
55. Fish published his figures as a note in an article by Phineas T. Whiting, "Heredity of Bristles in the Common Greenbottle Fly," *American Naturalist* 48 (1914): 343–44.

culture, where he continued work on genetic interaction systems affecting color inheritance. During 1917 and 1918 he published eleven papers on color inheritance in mammals. In addition to elucidating the interaction systems in color inheritance, Wright in these papers also used the Hardy-Weinberg equilibrium principle to calculate genotype frequencies in populations as a means of discriminating between genetical hypotheses.[56] At the time Wright wrote these papers he was not acquainted with the genetical work of Hardy or Weinberg. But the equilibrium principle which later carried both their names seemed intuitively obvious to Wright, and he referred to it as "the well known formula for a Mendelian population in equilibrium." [57]

At the Animal Husbandry Division, Wright also assumed charge of an extensive inbreeding experiment on guinea pigs begun in 1906 by G. M. Rommel. Faced with the task of analyzing the accumulated data, he became seriously interested in constructing a general mathematical theory of inbreeding. Seeing that the method previously utilized by Pearl, Jennings, Wentworth and Remick, and Robbins was too cumbersome with complex systems of mating, he searched for a new way to approach the problem of inbreeding. In 1920 he discovered that his method of path coefficients, which he had previously developed for other reasons, provided the powerful tool he needed for the analysis of systems of breeding in general.

Wright first used the mehod of path coefficients, although he did not use that terminology, in a paper in 1917 with the title "On the Nature of Size Factors." [58] In 1914, Castle had published a short paper on the correlations between five bone measurements in a stock of rabbits. He found that the correlations were "all positive and fairly high. . . . In view of the high correlations obtaining between one skeletal dimension and another . . . it follows that to a large extent the

56. Using the equilibrium principle Wright distinguished between a one-factor and a two-factor hypothesis in the coat color of cattle in "Color Inheritance in Mammals. 6. Cattle," *Journal of Heredity* 8 (1917): 521–27. He also showed that eye color in humans did not depend upon single factor inheritance in "Color Inheritance in Mammals. 11. Man," ibid., 9 (1918): 231–32.

57. Wright, "Color Inheritance in Mammals. 6. Cattle," p. 522.

58. *Genetics* 3 (1918): 367–74.

factors which determine size are *general* factors affecting all parts of the skeleton simultaneously."[59] Wright helped compute the correlation tables for Castle's paper. Then in 1917 Charles B. Davenport published a long paper on the inheritance of stature in man.[60] He claimed that size factors which affected only particular segments of stature were more important than factors which affected the growth of the body as a whole. Wright decided to reevaluate Castle's data in the light of Davenport's contentions. In his paper Wright proposed "to illustrate a method of analysis as well as to bring out certain conclusions."[61]

Wright's method was designed to estimate the degree to which a given effect was determined by each of a number of causes. In this case he treated the causes, namely, factors which affected general size and factors which affected separate parts, as independent. Wright's first formulation of the theory of path coefficients was as follows:

Let X and Y be two characters whose variations are determined in part by certain causes A, B, C, etc., which act on both and in part by causes which apply to only one or the other, M and N respectively. These causes are assumed to be independent of each other. Represent by small letters a, b, c, etc., the proportions of the variation of X determined by these causes and by a^1, b^1, c^1, etc., the proportions in the case of Y. The extent to which a cause determines the variation in an effect is measured by the proportion of the squared standard deviation of the latter for which it is responsible. This follows from the proposition that the squared standard deviations due to single causes acting alone may be combined by simple addition to find the squared standard deviation of an array in which all causes are acting, provided the causes are independent of each other. . . . As a, b, etc., are the proportions of the variation of X which are determined by the various causes

$$[I] \quad \begin{aligned} a + b + c + d &\dots\dots\dots\dots + m = 1 \\ a^1 + b^1 + c^1 + d^1 &\dots\dots\dots + n^1 = 1 \end{aligned}$$

59. W. E. Castle, *The Nature of Size Factors as Indicated by a Study of Correlation,* Carnegie Institution of Washington Publication, no. 196 (Washington, D.C., 1914), p. 51.

60. Charles B. Davenport, "Inheritance of Stature," *Genetics* 2 (1917): 313–89.

61. Wright, "On the Nature of Size Factors," p. 367.

It is easy to demonstrate the following proposition in regard to the correlation between X and Y.

$$[\text{II}] \quad r_{xy} = \pm \sqrt{aa^1} \pm \sqrt{bb^1} \pm \sqrt{cc^1} \ldots .^{[62]}$$

By making the sums of the degrees of determination equal to unity (equations I) and expressing the known correlation in terms of the unknown degrees of determination (equation II), a series of simultaneous equations were formed and could be solved for the unknown effects of single causes. In the case of size factors in rabbits, Wright's model showed that "most differences between individuals are those which involve the size of the body as a whole." [63]

Wright soon found another application for his new method. He had observed that even after twenty generations of intense inbreeding a family of guinea pigs with a tricolor coat might yield offspring with very different amounts of each color. But because of the intense inbreeding the family should be nearly homozygous. Wright suspected that the variations in coat color were caused by environmental factors. He wanted a method to measure the relative importance of heredity and environmental factors which he could apply to the case of coat color in his stocks of guinea pigs.

Wright's paper on this topic was published in 1920. He began by defining a path coefficient as "the ratio of the variability of the effect to be found when all causes are constant except the one in question, the variability of which is kept unchanged, to the total variability. Variability is measured by the standard deviation." [64] In the earlier paper the proportions of variation in a character determined by causes A, B, C, etc., and denoted by a, b, c, etc., were the squares of the respective path coefficients. Wright extended his theoretical model to include a simple case where the causes were correlated instead of independent. He then applied the method of path coefficients to the problem of ascertaining the relative importance of

62. Ibid., pp. 370–71.
63. Ibid., pp. 373–74.
64. Sewall Wright, "The Relative Importance of Heredity and Environment in Determining the Piebald Pattern of Guinea-Pigs," *Proceedings of the National Academy of Sciences* 6 (1920): 329.

heredity and environment in determining the piebald pattern in guinea pigs. He found that in the highly inbred stocks heredity determined almost none of the variability and irregularity in development determined almost all of it. This was of course the expected result because the inbred stock was probably very nearly homozygous.

In January 1921 Wright published, under the title "Correlation and Causation,"[65] his general theory of path coefficients. He was now able to treat systems of independent and correlated causes, nonadditive factors, and nonlinear relations. He derived a general formula for the expression of the correlation between any two variables in terms of path coefficients: "the correlation between two variables is equal to the sum of the products of the chains of path coefficients along all of the paths by which they are connected."[66] In this paper he applied the method to factors which determined birth weight in guinea pigs 'and to factors which affected the rate of transpiration in plants. Wright was aware of the flexibility of his method of path coefficients and he applied it in many ways during the next ten years.

Perhaps the most striking application of the method of path coefficients was to the effects of various systems of mating. Under the general heading of "Systems of Mating,"[67] Wright published in 1921 a series of five papers in which he explored the biometric relation between parent and offspring, the effects of various systems of inbreeding on the genetic composition of a population, the effects of assortative mating, and the effects of selection. The method of path coefficients was much easier to apply to the problems of inbreeding than the direct method utilized by earlier workers. Wright was able to quickly corroborate earlier researches and then extend his method to systems of mating for which the direct method was far too cumbersome. The method provided an easy way to calculate the increase in the percentage of homozygosis in successive generations under various systems of inbreeding. In

65. *Journal of Agricultural Research* 20 (1921): 557–85.
66. Ibid., p. 568.
67. Sewall Wright, "Systems of Mating," *Genetics* 6 (1921): 111–78.

the years 1923–26 Wright also used the theory of path co-efficients to analyse the history of inbreeding in the development of shorthorn cattle.

Having developed a general theory of the quantitative consequences of inbreeding, Wright was now prepared to analyze in depth the extensive results of the inbreeding experiments with guinea pigs which had continued under his direction since 1915.[68] Wright's conclusions about these inbreeding experiments are crucial to an understanding of his later views on the process of evolution. The data showed that the brother-sister mating was carried on for over twenty generations in many of the families, and a control stock was maintained. The highly inbred stocks exhibited a general decline in all elements of vigor, including mortality at birth and between birth and weaning, the size of litter, the weight at various ages, the regularity in producing litters, and the resistance to tuberculosis. Several families, however, despite intense inbreeding, exhibited no obvious degeneration.

One important effect of continued intense inbreeding was differentiation among the families. Fixation of many combinations of color, number of toes, elements of vigor, and various abnormalities occurred in each inbred family. When highly inbred stocks with fixed heritable characters were crossed there was a marked recovery of the vigor exhibited by the control stock.

Wright interpreted the results of the inbreeding experiments in accordance with the theories of East and Jones:[69]

The fundamental effect of inbreeding is the automatic increase in homozygosis in all respects. An average decline in vigor is the consequence of the observed fact that recessive factors, more extensively brought into expression by an increase in homozygosis, are more likely to be deleterious than are their dominant allelomorphs. The differentiation among the families is due to the chance fixation of different com-

68. Sewall Wright, "The Effects of Inbreeding and Crossbreeding on Guinea Pigs," *Bulletin of the U.S. Department of Agriculture,* nos. 1090 and 1121 (1922).
69. See Edward M. East and Donald F. Jones, *Inbreeding and Outbreeding* (Philadelphia: J. B. Lippincott, 1919).

binations of the factors present in the original heterozygous stock. Crossing results in improvement because each family in general supplies some dominant factors lacking in the others.[70]

Wright suggested a theory, based upon the experimental results obtained with guinea pigs, of how to combine inbreeding, crossbreeding, and selection for the most effective improvement of livestock. The characteristics he had studied in guinea pigs,

like most of those of economic importance with livestock, are of a kind which is determined only to a slight extent by heredity in the individual. . . . Progress by ordinary selection of individuals would thus be very slow or nil. A single unfortunate selection of a sire, good as an individual, but inferior in heredity, is likely at any time to undo all past progress. On the other hand, by starting a large number of inbred lines, important hereditary differences in these respects are brought clearly to light and fixed. Crosses among these lines ought to give a full recovery of whatever vigor has been lost by inbreeding, and particular crosses may safely be expected to show a combination of desired characters distinctly superior to the original stock. Thus a crossbred stock can be developed which can be maintained at a higher level than the original stock, a level which could not have been reached by selection alone.[71]

Wright was convinced by ten years of experimental work that interaction systems of genes were an important part of the genetic constitution of organisms. The advantage of his theory of artificial selection was that whole interaction systems would be fixed by inbreeding and could then be selected. Selection, instead of operating upon single gene effects, could operate upon entire interaction systems. Wright believed that simple direct selection for single gene effects was far less effective than the selection of interaction systems.

He soon began to apply his conclusions about effective livestock breeding to the problem of evolution in nature. By 1925, when he left the Animal Husbandry Division for a

70. Wright, "Effects of Inbreeding and Crossbreeding on Guinea Pigs," no. 1121, pp. 48–49.
71. Ibid., p. 49.

position at the University of Chicago, Wright had written a long paper on evolution. He was dissatisfied with some of his mathematical calculations and withheld the paper from publication until 1931, but his general approach to the problem of evolution was clear in his mind by 1925.

Wright was convinced from his experimental work that interaction systems were important in evolution and that the random drift of genes caused by inbreeding was important for the creation of novel interaction systems. His theory of evolution was constructed with these ideas in mind. Thus from the beginning of his work on evolution Wright differed markedly from Fisher, who denied the importance of genic interaction and random genetic drift in evolution. Fisher believed that natural selection operated most effectively in large populations, because more variant genes were stored there. Natural selection, as stated by his fundamental theorem, acted to increase the fitness of a single interaction system by single gene replacements. Wright on the other hand believed that natural selection operated most effectively in smaller populations where inbreeding was sufficiently intense to create new interaction systems through random drift but not intense enough to cause random nonadaptive fixation of genes. Natural selection could then act upon the new interaction systems. In this way the population could change much more rapidly than by mass selection of single genes.

Wright disagreed with Fisher's theory of dominance. In his first paper on dominance Fisher had stated "that with mutation rates of one in a million the corresponding selection in the state of nature, though extremely slow, can not safely be neglected in the case of the heterozygotes." [72] Using Fisher's hypothesis, Wright calculated the selection pressures operating upon modifiers of dominance in heterozygotes and found they were of the order of mutation rates. Because of his strong belief in the universality of genic interaction, he doubted that such small selection pressures were important in the fixation of modifiers of dominance: "It has been shown

72. Fisher, "Possible Modification of the Response of the Wild Type to Recurrent Mutations," p. 126.

that genes often have multiple effects and it is not unlikely
. . . that in general any given gene has some effect on nearly
all parts of the organism." [73] Thus when selection acted directly upon some parts of an organism, it acted indirectly on
others. Direct selection pressures upon a character determined
by gene A could cause selection pressure on selectively neu-
tral characters determined by genes which interacted with
gene A. Wright believed these indirect selection pressures
were generally greater than the order of mutation pressure.

It will be seen that the hypothesis that a selection pressure
[of the order of mutation rates] can be the *general* factor making for dominance of wild type, depends upon the assumption
that modifiers of dominance (assumed to be sufficiently abundant) are in general so nearly indifferent to selection on their
own account that a force of the order of mutation pressure is
the *major* factor controlling their fate. With the prevalence of
multiple effects in mind it seems doubtful to the present writer
whether there are many such genes.[74]

Having rejected Fisher's explanation of the evolution of
dominance, Wright proposed an alternative explanation in
accordance with his experience with genic interaction systems affecting coat color in guinea pigs:

Probably most geneticists would hold that dominance in
general has some immediate physiological explanation. Bateson long ago suggested that pairs of allelomorphs represent the
presence or absence of something and that it was to be expected that one dose of an entity would give a result more
like that of two doses than like complete absence. There are
many reasons which have led to the general abandonment of
the presence and absence hypothesis in its literal form. There
is still much to be said, however, for the idea that the common-
est type of change in a gene is one which partially or completely inactivates it in one or more respects. . . . It seems
that in the hypothesis that mutations are most frequently in
the direction of inactivation and that for physiological reasons inactivation should generally behave as recessive, at least
among factors with major effects, may be found the explana-

73. Sewall Wright, "Fisher's Theory of Dominance," *American
Naturalist* 63 (1929): 276–77.
    74. Ibid., p. 277.

tion of the prevalence of recessiveness among observed mutations.[75]

Wright was here speaking of a single genetic background. In general he believed that "dominance is a phenomenon of the physiology of development to be associated with the various types of epistatic relationships among factors."[76] In one genetic background allele A_1 might be dominant to allele A_2, but in another genetic background A_2 might be dominant to A_1. Wright's view of dominance was really an extension of his general view of the importance of interaction effects, the same view which so deeply influenced his concept of evolution. Since Fisher was convinced that interaction effects were not important in evolution, he and Wright were never able to agree on the evolution of dominance despite continued communication in the journals.

By the time Fisher published his *Genetical Theory of Natural Selection* in 1930, Wright had put his paper on evolution in nature in nearly final form. Fisher and Wright had corresponded from 1928 on and each had pointed out mistakes in the work of the other. Thus when Wright wrote a review of the *Genetical Theory of Natural Selection* soon after its appearance, he was fully prepared to compare his own theory of evolution with that of Fisher.[77]

Instead of approaching the problem of the distribution of gene frequencies by statistical methods suitable for the analysis of variance in very large populations as Fisher had, Wright approached the problem through analysis of inbreeding by an application of his method of path coefficients. But as a result of the correspondence of 1928–30 Wright could say that "our mathematical results on the distribution of gene frequencies are now in complete agreement as far as comparable, although based upon very different methods of attack."[78] Despite this agreement Fisher and Wright differed markedly in their interpretations of the mathematical results.

75. Ibid., pp. 277–78.
76. Ibid., p. 274.
77. Sewall Wright, "The Genetical Theory of Natural Selection. A Review," *Journal of Heredity* 21 (1930): 349–56.
78. Ibid., p. 35.

Fisher believed that the mass selection of small single gene effects in large populations was the primary process of evolution. In this process the effects of small random fluctuations in the frequencies of genes tended to cancel each other and were therefore negligible. Wright thought the random genetic drift caused by inbreeding was actually very important in evolution. He stated that Fisher "overlooks the role of inbreeding as a factor leading to non-adaptative differentiation of local strains, through selection of which, adaptive evolution of the species as a whole may be brought about more effectively than through mass selection of individuals." [79]

Wright also disagreed with the great emphasis Fisher placed on his "fundamental theorem of natural selection" and believed it needed revision. The theorem, Wright said,

assumes that each gene is assigned a constant value, measuring its contribution to the character of the individual (here fitness) in such a way that the sums of the contributions of all genes will equal as closely as possible the actual measures of the character in the individuals of the population. Obviously there could be exact agreement in all cases only if dominance and epistatic relationships were completely lacking. Actually dominance is very common and with respect to such a character as fitness, it may safely be assumed that there are always important epistatic effects. Genes favorable in one combination, are, for example, extremely likely to be unfavorable in another. [80]

Having stated his disagreement with Fisher's interpretation of evolution, Wright outlined his own. It is important to quote Wright at some length because of the many erroneous interpretations of his view of evolution which have appeared in print, even recently. Some geneticists have associated with Wright the belief that random drift is the major factor in evolution and that selection is somehow of lesser importance. It is clear in what follows that Wright considered random drift important because it helps create the gene combinations upon which selection acts. The paragraphs quoted from Wright's papers also prove erroneous Ernst Mayr's

79. Ibid., p. 350.
80. Ibid., p. 353.

well-known accusation (made at the Cold Spring Harbor
Symposium of 1959) that Wright, along with Fisher and
Haldane, was a "beanbag" geneticist. The truth is that from
the beginning of his career in genetics Wright was interested
in interaction systems of genes, and his concept of evolution
in nature rested upon the belief that selection was most effective when acting upon interaction systems of genes rather
than upon single genes. Here then is Wright's vision of evolution in nature as he saw it in 1930:

> If the population is not too large, the effects of random
> sampling of gametes in each generation brings about a ran-
> dom drifting of the gene frequencies about their mean posi-
> tions of equilibrium. In such a population we can not speak
> of single equilibrium values but of probability arrays for each
> gene, even under constant external conditions. If the popula-
> tion is too small, this random drifting about leads inevitably
> to fixation of one or the other allelomorph, loss of variance,
> and degeneration. At a certain intermediate size of popula-
> tion, however (relative to prevailing mutation and selection
> rates), there will be a continuous kaleidoscopic shifting of the
> prevailing gene combinations, not adaptive itself, but provid-
> ing an opportunity for the occasional appearance of new adap-
> tive combinations of types which would never be reached by
> a direct selection process. There would follow thorough-
> going changes in the system of selection coefficients, changes
> in the probability arrays themselves of the various genes and
> in the long run an essentially irreversible adaptive advance
> of the species. It has seemed to me that the conditions for
> evolution would be more favorable here than in the indefi-
> nitely large population of Dr. Fisher's scheme. It would, how-
> ever, be very slow, even in terms of geologic time, since it
> can be shown to be limited by mutation rate. A much more
> favorable condition would be that of a large population, broken
> up into imperfectly isolated local strains. . . . The rate of evo-
> lutionary change depends primarily on the balance between
> the effective size of population in the local strain and the
> amount of interchange of individuals with the species as a
> whole and is therefore not limited by mutation rates. The
> consequence would seem to be a rapid differentiation of local
> strains, in itself non-adaptive, but permitting selective increase
> or decrease of the numbers in different strains and thus lead-

ing to relatively rapid adaptive advance of the species as a whole.[81]

This concept of the evolutionary process Wright later termed the "three-phase shifting balance" theory, involving random drift, intrademe, and interdeme selection. When Wright's long paper, entitled "Evolution in Mendelian Populations," [82] appeared in 1931, it not only provided corroboration of Fisher's earlier published mathematical considerations by a different method but also provided a significantly different interpretation of the evolutionary process as a whole. In one basic way Wright's efforts resembled those of Fisher. Wright stated in the introduction to his paper that

the rediscovery of Mendelian heredity in 1900 came as a direct consequence of de Vries' investigation. Major Mendelian differences were naturally the first to attract attention. It is not therefore surprising that the phenomena of Mendelian heredity were looked upon as confirming de Vries' theory. . . . Johannsen's study of pure lines was interpreted as meaning that Darwin's selection of small random variations was not a true evolutionary factor.[83]

Wright considered his own work to be a culmination of the reaction to this point of view.

J. B. S. HALDANE

John Burdon Sanderson Haldane was born on 5 November 1892. His father was the physiologist John Scott Haldane. The Haldane family provided an extraordinarily stimulating environment for young J. B. S.[84] His intellectual curiosity was encouraged by his parents in many ways. For example, in 1901 his father took eight-year-old J. B. S. to a lecture by A. D. Darbishire on the recently discovered work of Mendel. Young Haldane was impressed.

81. Ibid., pp. 354–55.
82. *Genetics* 16 (1931): 97–159.
83. Ibid., p. 99.
84. For an account of Haldane's childhood, see Ronald W. Clark, *JBS: The Life and Work of J. B. S. Haldane* (New York: Coward-McCann, 1969), part 1.

Haldane attended Eaton, then entered New College, Oxford, in 1911 on a mathematics scholarship. He excelled in mathematics, gaining first-class honors within a year. Then he switched from mathematics to "Greats," primarily classics and philosophy. In 1914 he won a First in "Greats." He never took a scientific degree. While at New College he attended E. S. Goodrich's biology lectures and found them stimulating. This was his only formal training in biology.

Haldane's practical interest in genetics was aroused even before he went to New College. His sister Naomi began raising guinea pigs for fun in 1908. He and Naomi conducted experiments with them, looking among other things for evidence of Mendelian inheritance. Haldane began reading the available literature on Mendelism, including Darbishire's papers on heredity in mice. From a study of Darbishire's data Haldane believed he had found evidence of linkage. In 1912 he presented this view before a seminar organized by Goodrich. He considered publishing the paper and wrote for advice to Punnett, who advised him to obtain his own data. Together with Naomi and a fellow student, A. D. Sprunt, Haldane began breeding experiments with mice and rats. Sprunt was killed in France early in World War I and Haldane, who was a lieutenant in the Black Watch, decided to publish a preliminary report on the research work in case he also was killed. The report, published in the *Journal of Genetics* in 1915, strongly indicated the existence of linkage in mice.[85] This was one of the first published cases of linkage in mammals.

Haldane was impressed by the work of Morgan and his colleagues on *Drosophila,* and although unable to conduct his researches during the war, he did keep up with the *Drosophila* work by reading journals while stationed in New Delhi during 1917 and 1918. He was especially interested in the problems of linkage and chromosome mapping. The *Drosophila* workers had made chromosome maps by recording frequencies of crossing over between genes on a chromo-

85. J. B. S. Haldane, A. D. Sprunt, and N. M. Haldane, "Reduplication in Mice," *Journal of Genetics* 5 (1915): 133–35.

some and making the chromosome distance between any two genes proportional to the frequency of crossing over between them. But it was well known that the measured distance between two distant loci on a chromosome map was less than the sum of the distances between loci located between the original two. Haldane began research on the linkage problem immediately after the war and published two papers on the subject in 1919. In the first he derived formulas for the probable errors of calculated linkage values.[86] He hoped geneticists would check their linkage data with the probable errors to see if the theory of crossing over (and chromosome mapping) was actually supported by the data. In the second paper he developed a theory to correct the discrepancies of chromosome maps based only upon linkage values. His theory permitted "the calculation of one of the cross-over values for three factors from the other two, with a probable error of less than 2%" and could also be used "to calculate the total length of a chromosome, and the number of double and triple crossovers to be expected in a large distance."[87] Haldane found his theory fit plant data accurately but the *Drosophila* data less well. The method used by Haldane in these papers was typical of his life-long approach to genetical problems. He discovered problems in the data of others for which he offered theoretical solutions, then checked his ideas with the data of others.

In 1922 Haldane spoke with the mathematician H. T. J. Norton of Trinity College, Cambridge. Norton had prepared the selection table in Punnett's *Mimicry in Butterflies* (1915). Haldane discovered that with the exception of Warren's short paper in 1917, Norton's table was still the only available analysis of the mathematical consequences of selection. He decided to undertake this investigation. By the time his book *The Causes of Evolution* appeared in 1932, he had published nine

86. J. B. S. Haldane, "The Probable Errors of Calculated Linkage Values, and the Most Accurate Method of Determining Gametic from Certain Zygotic Series," *Journal of Genetics* 8 (1919): 291–97.

87. J. B. S. Haldane, "The Combination of Linkage Values, and the Calculation of Distance between the Loci of Linked Factors," *Journal of Genetics* 8 (1919): 308, 309.

papers under the collective title "A Mathematical Theory of Natural and Artificial Selection."

Haldane published the first paper in the series in 1924. He opened with the statement:

A satisfactory theory of natural selection must be quantitative. In order to establish the view that natural selection is capable of accounting for the known facts of evolution we must show not only that it can cause a species to change, but that it can cause it to change at a rate which will account for present and past transmutations.[88]

The purpose of his series of papers was to quantify the theory of natural selection. In the first paper Haldane derived mathematical expressions and computed tables for the effect of selection on simple Mendelian populations. He assumed random mating, an infinite population, separate generations, complete dominance, perfect Mendelian segregation, and no change of selection intensities from generation to generation. The cases he treated varied from selection in self-fertilizing populations, to selection of dominant or recessive autosomal and sex-linked characters, to prenatal selection. The method he used was to derive recurrence equations from which the proportion of gametes in one generation could be computed from the proportion in the preceding generation. Most of these recurrence equations were nonlinear and Haldane had to work out approximate solutions.

He applied the model to the case of the peppered moth *Amphidasys betularia*. A dominant melanic form had first appeared in this variety at Manchester in 1848 and before 1901 it had replaced the recessive form. Haldane found that "the fertility of the dominants must be 50% greater than that of the recessives," which he called a "not very intense degree of natural selection." [89] But it was far more intense than the selection pressures which Fisher believed were important in the evolution of species. Haldane, like Morgan, placed much

88. J. B. S. Haldane, "A Mathematical Theory of Natural and Artificial Selection. Part 1," *Transactions of the Cambridge Philosophical Society* 22 (1924): 19.
89. Ibid., p. 26.

more emphasis upon the selective importance of a single gene effect than did Fisher.

In succeeding parts of the series Haldane modified his initial assumptions, then explored the mathematical consequences of selection. In part 2 (1924) [90] he treated selection with partial self-fertilization, inbreeding, assortative mating, and selective fertilization. Part 3 (1926) [91] covered the problems of selection with incomplete dominance. Part 4 (1927) [92] was devoted to a study of the consequences of selection when generations were overlapping instead of separate. The finite difference equation Haldane had previously developed was inapplicable, and he here developed another finite difference equation for the change in the proportions of genotypes between times t_1 and t_2. In part 5 (1927) [93] he treated the problem of the survival of new mutations, dominant or recessive, and examined the balance between mutation pressure and selection pressure. He found, as had Fisher, that the survival of very rare recessive mutations was a stochastic process which depended little on the deterministic effects of selection.

There was a three-year gap, from 1927 to 1930, between the appearance of part 5 and part 6. During this time Haldane wrote several articles on the process of evolution in general and joined in the dispute between Fisher and Wright regarding the evolution of dominance. By 1927 Haldane's general approach to evolution was settled. As a young man he was interested in single gene effects, probably because of the influence of the English Mendelians and the *Drosophila* workers. He believed that selection was the most powerful force in evolutionary change and found Morgan's general view of evolution by single gene replacement appealing. Up to 1927 Haldane treated populations as very large or infinite and selection as completely deterministic, except for the fixation of

90. J. B. S. Haldane, "A Mathematical Theory of Natural and Artificial Selection. Part 2," *Proceedings of the Cambridge Philosophical Society. Biological Sciences* 1 (1924): 158–63.

91. J. B. S. Haldane, "A Mathematical Theory of Natural and Artificial Selection. Part 3," *Proceedings of the Cambridge Philosophical Society* 23 (1926): 363–72.

92. Ibid., pp. 607–15.

93. Ibid., pp. 838–44.

rare recessives which depended upon a stochastic process. Un-like Fisher, he emphasized the importance of high selection pressures caused by single mutations. But he agreed with Fisher about the deterministic effect of mass selection and the importance of this process in evolution. Wright believed that the most effective population size was much smaller than did either Haldane or Fisher and that the effect of selection on single genes was less important than the effect of selection upon interaction systems of genes.

Haldane thought his investigations exemplified the possibilities of Darwinian selection. Like Fisher and Wright, he found it necessary to defend the principle of natural selection even as late as 1929:

> Quantitative work shows clearly that natural selection is a reality, and that, among other things, it selects Mendelian genes, which are known to be distributed at random through wild populations, and to follow the laws of chance in their distribution to offspring. In other words, they are an agency producing variation of the kind which Darwin postulated as the raw material upon which selection acts.[94]

Later in 1929 Haldane wrote that he believed in the importance of chromosomal alterations and polyploidy as factors in evolution.[95] In this he tended to agree with Wright rather than Fisher and the Darwinian tradition.

Early in 1930 Haldane entered the controversy over the evolution of dominance. First he argued in agreement with Wright that not enough modifiers of the sort postulated by Fisher were available in natural populations. Then he proposed an alternative hypothesis:

> Adopting Goldschmidt's view that genes are catalysts acting at a definite rate, there is no obvious way of distinguishing those which act at more than a certain rate. E.g., if an enzyme can oxidize a certain substance as quickly as it is formed, no visible result arises from doubling the amount of that enzyme. Hence, while a minus mutation (diminution of activity) of a normal gene may yield a recessive type, a plus

94. J. B. S. Haldane, "Natural Selection," *Nature* 124 (1929): 444.
95. J. B. S. Haldane, "The Species Problem in the Light of Genetics," ibid., pp. 514–16.

mutation is often unobservable. Now on this hypothesis we have to explain why a wild-type gene generally has a factor of safety of at least 2, as is shown by the fact that one wild-type gene has nearly the same effect as two. If we imagine a race whose genes were only just doing the work required of them, then any inactivation of one of a pair of genes would lead to a loss of total activity. Thus if A_1A_1 can just oxidize all of a certain substrate as fast as it is formed, its inactivation will produce a zygote A_1a which can only oxidize about half. If now A_1 mutates A_2, which can oxidize at twice or thrice the rate of A_1, if necessary, no effect will be produced, i.e., A_1A_2 and A_2A_2 zygotes will be indistinguishable from A_1A_1. But A_2a will be normal. Hence A_2a zygotes will have a better chance of survival than A_1a, and A_2 will be selected.

In other words the modifiers postulated by Fisher are probably the normal allelomorphs of mutant genes, and the Fisher effect is rather to accentuate the activity of genes already present than to call up new modifiers.[96]

Haldane thus solved the problem of the evolution of dominance by means of the selection of sizable mutations, in accordance with his general view of evolution by single gene replacement.

Between 1930 and 1932 Haldane published four more parts to his "Mathematical Theory of Natural and Artificial Selection." Part 6 (1930)[97] dealt with the effects of isolation and migration. In part 7 (1931)[98] he finally explored a case where selection intensity was not constant from generation to generation, specifically when selection intensity was a function of mortality rate. In part 8 (1931)[99] he started with the supposition that a population with interacting genes was very close to genetic equilibrium (which Haldane termed "metastable equilibrium"). He then considered the case where two genes might be deleterious singly but advantageous together and outlined the most favorable conditions for the replacement of one interaction system by another. He concluded that "in

96. J. B. S. Haldane, "A Note on Fisher's Theory of the Origin of Dominance, and on a Correlation between Dominance and Linkage," *American Naturalist* 64 (1930): 88.

97. *Proceedings of the Cambridge Philosophical Society* 26 (1930): 220–30.

98. Ibid. 27 (1931): 131–36.

99. Ibid., pp. 137–42.

many cases related species represented stable types . . . and
the process of species formation may be a rupture of the
metastable equilibrium. Clearly such a rupture will be specially likely where small communities are isolated." [100] Haldane submitted this paper on 20 November 1930 before he
could have read Wright's "Evolution in Mendelian Populations." Thus he independently derived results in agreement
with those Wright had derived earlier but had not yet published in substantial form. But Haldane believed that small
partially isolated communities were far less common in natural populations than did Wright. In part 9 (1932),[101] the last
to appear before the publication of *The Causes of Evolution,*
Haldane treated the effects of rapid selection, a process which
Fisher considered relatively unimportant in the evolution of
natural populations.

In January 1931 Haldane delivered a series of lectures entitled "A Re-examination of Darwinism." The series was
published as a book in 1932 with the title *The Causes of Evolution.*[102] The book contained an exposition of his general
view of evolution. He intended, as had Fisher in the *Genetical Theory of Natural Selection* and Wright in "Evolution in
Mendelian Populations," to dispel the belief that Mendelism
had killed Darwinism. He headed the first chapter with the
quote "Darwinism is dead." He raised most of the charges
brought against Darwinian selection between 1900 and 1930
and, while emphasizing his own view that "natural selection
is an important cause of evolution," concluded that "the criticism of Darwinism has been so thorough-going that a few
biologists and many laymen regard it as more or less ex-
ploded." [103]

Included in *The Causes of Evolution* was a technical appendix summarizing the most important points developed in
the first nine parts of "A Mathematical Theory of Natural
and Artificial Selection." Haldane also analyzed the contri-

100. Ibid., p. 141.
101. Ibid., 28 (1932): 244–48.
102. (London: Longmans, Green, 1932).
103. Ibid., pp. 20, 32.

butions of Fisher and Wright in the appendix. He agreed with much of Fisher's work and relied heavily upon *The Genetical Theory of Natural Selection* in writing the appendix. The basic issues on which Haldane differed from Fisher concerned the evolution of dominance and the intensity of selection pressure caused by a single gene effect in a natural population. Less basic differences were Haldane's greater emphasis upon migration, genic interaction, and discontinuities in evolution. Haldane agreed with Wright's view that evolution should be slow in very small populations but disagreed that evolution was also slow in very large populations. He thought Wright overemphasized the importance of random genetic drift. But he agreed with one of Wright's basic positions:

Wright's theory certainly supports the view taken in this book that the evolution in large random-mating populations, which is recorded by paleontology, is not representative of evolution in general, and perhaps gives a false impression of the events occurring in less numerous species. It is a striking fact that none of the extinct species, which, from the abundance of their fossil remains, are well known to us, appear to have been in our own ancestral line. Our ancestors were mostly rather rare creatures.[104]

After discussing the views of Fisher and Wright as well as his own, Haldane closed the *Causes of Evolution* with the statement:

The permeation of biology by mathematics is only beginning, but unless the history of science is an inadequate guide, it will continue, and the investigations here summarized represent the beginning of a new branch of applied mathematics.[105]

In the light of modern population genetics, this statement was indeed prophetic.

Having examined the views of evolution of Fisher, Haldane, and Wright, it is illuminating to compare the general view each took of the work of the other two in the early

104. Ibid., pp. 213–14.
105. Ibid., p. 215.

1930s. Haldane stated that the work of Wright "resembles the work of Fisher more than that of Haldane." [106] Wright believed that the strong emphasis upon the deterministic effects of mass selection of single genes in the work of Fisher and Haldane distinguished their work sharply from his own, which emphasized the selection of interaction systems of genes. Fisher thought Wright and Haldane failed to appreciate the importance of very small selection pressures acting over long periods of time in the evolution of natural populations. He often lumped Wright and Haldane together as critics of his views. Thus the relationship between the three appears to have been symmetrical.

The work of Fisher, Haldane, and Wright up to 1932 is the culmination of this account. It began with the disagreement which arose immediately upon the publication of Darwin's *Origin of Species* between the adherents of continuous evolution and the adherents of discontinuous evolution. Even Darwin's friends Huxley and Galton opposed his view of the continuity of evolution. The idea of discontinuous evolution was reinforced by the work of William Bateson in the 1890s. His work stimulated heated opposition from the biometricians, who believed in Darwin's idea of selection by minute differences as the mechanism of evolution. When Mendelian heredity was rediscovered in 1900 Bateson grabbed it as a support for discontinuous evolution and the biometricians reacted by attacking Mendelism, with the result that in the first decade of this century Mendelism and Darwinism were generally thought to be contradictory. The consequences inferred from the pure line theory helped to further the split between Mendelian heredity and Darwinian selection. Many Mendelians thought the pure line theory necessarily led to de Vries's mutation theory rather than to Darwin's idea of continuous evolution. But between 1908 and 1918 crucial selection experiments demonstrated that selection of small differences could change a population significantly and permanently. By 1918 many prominent geneticists were promulgating the view that

106. Ibid., p. 212.

Mendelian heredity and Darwinian selection were complementary rather than contradictory. Fisher, Haldane, and Wright then quantitatively synthesized Mendelian heredity and natural selection into the science of population genetics. The work of all three was to some extent a reaction against Mendelians who claimed that natural selection was of subordinate importance in evolution.

CONCLUSIONS

The story of the origins of population genetics illustrates three important patterns in the history of science, all of which contradict the current popular conception of science. First, it illustrates that personality conflicts are sometimes very important in the development of scientific ideas. The intense antagonisms generated by Bateson's dislike of Pearson and Weldon and vice versa contributed to a delay of more than a decade in the understanding that Mendelism and Darwinism were complementary. If Bateson and Pearson had collaborated instead of fought, population genetics would have gained a significantly earlier start. Second, the story illustrates that the acceptance by scientists of a new idea is sometimes more dependent upon its a priori acceptability than upon its scientific proof. The pure line theory is an example. It was accepted, as well as the selection theory associated with it, by almost all geneticists even though conclusive experimental proof was totally absent. Third, the story of the origins of population genetics illustrates that a field of science can begin with a theoretical structure which is far from consistent. Population genetics was founded by three men, each of whom produced a basic model of evolutionary change. All three agreed upon the importance of natural selection, but each had a significantly different approach. The problem was that each could cite examples from natural populations to support his approach. Population genetics has grown enormously and attracted much attention since its inception, but population geneticists have yet to remove from the theoretical framework many of the basic differences of approach already visible in 1932 in the work of Fisher, Haldane, and Wright. For

example, each proposed a model for the evolution of dominance and these have yet to be authoritatively reconciled. This
situation results because small parameters can effect great
changes in populations over a surprisingly small number of
generations. But the isolation of all significant parameters affecting population change is difficult even under the best conditions with populations in the laboratory. With populations
in nature this problem of course greatly increases. Thus with
the gap between theoretical models and available observational data so large, population genetics began and continues
with a theoretical structure containing obvious internal inconsistencies.

APPENDIX   Galton, Pearson, and the Law of Ancestral Heredity

THE CONFUSION OF BIOLOGISTS CONCERNING THE MEANING AND application of Galton's law of ancestral heredity will perhaps be more understandable if the various interpretations given it by Galton and Pearson are distinguished.[1] Galton contributed to the confusion over his law in three ways.

First, Galton's forms A and B of his law are mathematically inconsistent. Consider the following situation. Suppose a mid-parent has a deviation of D_1 from the mean of the population. Then the offspring will have an average deviation of $D = \frac{2}{3} D_1$ by the law of regression. Galton's form A says the mid-parent contributes one half of the heritage of the offspring, or $\frac{1}{2}(\frac{2}{3} D_1)$, the mid-grandparent contributes $\frac{1}{4}(\frac{2}{3} D_1)$, etc. Then the total contribution of the ancestors to the deviation of the offspring is

$$D = \frac{1}{2}(\frac{2}{3} D_1) + \frac{1}{4}(\frac{2}{3} D_1) + \frac{1}{8}(\frac{2}{3} D_1) + \cdots = \frac{2}{3} D_1.$$

In this formulation the series $\frac{1}{2}$, $\frac{1}{4}$, $\frac{1}{8}$ ... appeals to Galton because it sums to 1 and accounts for the entire heritage of the offspring. But form B of Galton's law states that the mid-parent contributes one half of its *own* deviation, or $\frac{1}{2} D_1$, and that the mid-grandparent contributes $\frac{1}{4} D_2$, etc., where D_1, D_2, D_3, etc., are the deviations of the individual ancestral generations. Form B of Galton's law thus generates the following series:

$$D = \frac{1}{2} D_1 + \frac{1}{4} D_2 + \frac{1}{8} D_3 + \cdots = \frac{2}{3} D_1.$$

The problem is that the "contribution" of any given ancestral generation is different acccording to whether one uses Galton's form A or form B. In form A the contribution of the mid-parent to the deviation of the offspring is $\frac{1}{2}(\frac{2}{3} D_1)$, or $\frac{1}{3} D_1$; in form B the contribution of the mid-parent to the

---

1. The text in chapter 2 dealing with this topic should be read before this Appendix. Some details mentioned there are assumed here.

deviation of the offspring is $\frac{1}{2} D_1$. Galton used forms A and B of his law as if they were interchangeable, contributing to the confusion concerning his law.

Second, Galton argued that his statistical law of ancestral heredity was based upon a physiological law which was a consequence of his "stirp" theory of heredity. In his 1897 article on the inheritance of coat color in Basset hounds, Galton, after stating his law in form B,

$$M + D = \frac{1}{2}(M + D_1) + \frac{1}{4}(M + D_2) + \text{etc.} = M + (\frac{1}{2} D_1 + \frac{1}{4} D_2 + \text{etc.}),$$

went on to say:

> It should be noted that nothing in this statistical law contradicts the generally accepted view that the chief, if not the sole, line of descent runs from germ to germ and not from person to person. The person may be accepted on the whole as a fair representative of the germ, and, being so, the statistical laws which apply to the persons would apply to the germs also, though with less precision in individual cases. Now this law is strictly consonant with the observed binary subdivisions of the germ cells, and the concomitant extrusion and loss of one-half of the several contributions from each of the two parents to the germ-cell of the offspring. The apparent artificiality of the law ceases on these grounds to afford cause for doubt; its close agreement with physiological phenomena ought to give a prejudice in *favour* of its truth rather than the contrary.[2]

Without stating it in quantitative form Galton had actually propounded a law of physiology which supported his law of ancestral contributions. He argued as follows: The individual is a fair representative of the germ plasm. There is continuity of the germ plasm from generation to generation. But each parent can transmit only one half of its germ plasm or else the offspring would have twice the amount of germ plasm as an individual parent. Thus the law of physiology may be stated: *the germ plasm of an individual contains contributions from all its ancestors, the amount of the contribution being larger as the ancestor is nearer.* On the basis of these physiological

2. Galton, "The Average Contribution of Each Several Ancestor to the Total Heritage of the Offspring," p. 403.

considerations and others connected with the series ½, ¼, ⅛ . . . , Galton suggested his law of ancestral contributions "might be inferred with considerable assurance *a priori.*" [3]

But within five years the Mendelians had found that in an experimental population with much heterozygosity an individual might have hidden recessives and was therefore not a fair representative of the germ plasm. Furthermore, the Mendelians believed that by halving the germ plasm the phenotypic characters were not on the average halved, as Galton implied. Instead they believed that some characters would disappear in toto and others would remain in toto. Having disproved Galton's law of physiology, most Mendelians also rejected Galton's law of ancestral heredity. But the situation was not that simple. By 1900, when Mendelian heredity was rediscovered, Karl Pearson had moved the law of ancestral heredity to a purely phenotypic level independent of whatever physiological mechanism of heredity might be operating.

A third element of confusion in Galton's thinking concerned the kind of variation to which his law of ancestral heredity applied. He had said in *Natural Inheritance* and later publications that the law did not apply to sports, the variation he considered important for evolutionary change. But he believed the law was applicable to "alternative" inheritance, as in the cases of coat color in Basset hounds and eye color in humans. Galton apparently considered the difference between blue and brown eyes to be within the realm of normal variation in a variety and not to be the result of sporting.

When the law of ancestral heredity was challenged by botanists who cited evidence from plant hybridizations, Galton replied:

Permit me to take this opportunity of removing a possible misapprehension concerning the scope of my theory. That theory is intended to apply only to the offspring of parents who, being of the *same variety,* differ in having a greater or less amount of such characteristics as any individual of that variety may normally possess. It does *not* relate to the off-

3. Ibid.

spring of parents of different varieties; in short it has nothing
to do with hybridism, for in that case the offspring of two
diverse parents do not necessarily assume an intermediate
form.[4]

It is difficult to reconcile Galton's statement here with his
expressed belief that his law of ancestral heredity applied to
coat color in Basset hounds, a characteristic which does not
blend. It is clear, however, that Galton wished to make a
distinction between the normal differences within a single
variety and the differences between varieties. His law of
ancestral heredity applied only to the former.

Problems with Galton's distinction between inter- and
intravarietal differences arose with the rediscovery of Mendelian heredity. Mendel's experiments were, as the title of
his paper says, in plant hybridizations. Thus supporters of
Galton's law stated it could not be challenged by experiments
with Mendelian heredity. But the Mendelians soon found
that intravarietal differences might also behave as Mendelian
characters. It was well known that double recessives, even
though the offspring of heterozygous parents of very different
appearance, would breed true indefinitely. This contradicted
Galton's law of ancestral heredity and most Mendelians concluded it must be false. Pearson escaped from this dilemma by
removing the application of Galton's law from alternative
inheritance. Thus until blended inheritance was adequately
interpreted in terms of Mendelian heredity (see chap. 4), Pearson's interpretation of Galton's law was safe from the onslaughts of the Mendelians.

Galton's quantitative law of "the average contribution of
each several ancestor to the total heritage of the offspring"
fascinated Pearson from the beginning. When he read *Natural Inheritance,* however, he had been inclined to dismiss
Galton's law because of the inadequate way it had been derived. At that time (1889), and in 1896 when he published a

4. Francis Galton, letter to *Nature,* 12 October 1897, cited by
Pearson, *Galton,* 3:41.

long theoretical paper on heredity and evolution,[5] Pearson accepted Galton's belief that if the correlation between one parent and offspring was r, then the correlation between one grandparent and offspring was r^2, etc.

By 1898 Pearson had changed his mind. In his paper on the law of ancestral heredity he stated that Galton's law and the ancestral correlation series r, r^2, r^3, . . . were inconsistent. And although he had earlier accepted the correlation series and rejected the general law, he now believed that the general law carried much weight and that the correlation series must be rejected. Indeed, Pearson now believed that Galton's law was an important new principle. Galton's law, he said:

if properly interpreted, . . . enables us to predict *a priori* the values of all the correlation coefficients of heredity, and forms, I venture to think, the fundamental principle of heredity from which all the numerical data of inheritance can in the future be deduced, at any rate, to a first approximation.[6]

This statement reveals the optimism of a mathematician looking at biology without the aid of firsthand acquaintance with biological organisms. Pearson's decision to name the law of ancestral heredity after Galton was meant as a tribute to Galton. But the law, "properly interpreted," as Pearson put it, was much different from Galton's own formulation.

In the 1897 paper on Basset hounds, Galton expressed his law in two forms, A and B, as mentioned above. Pearson never utilized form A. He began with form B,

$$M + D = \tfrac{1}{2}(M + D_1) + \tfrac{1}{4}(M + D_2) + \text{etc.} = M + (\tfrac{1}{2}D_1 + \tfrac{1}{4}D_2 + \text{etc.}),$$

then changed many of the assumptions Galton had made in this formulation. Pearson rejected the assumptions that the mean M of the population stayed constant generation after generation and that the variability, as measured by the stand-

5. Karl Pearson, "Mathematical Contributions to the Theory of Evolution. 3. Regression, Heredity, and Panmixia," *Philosophical Transactions of the Royal Society*, A, 187 (1896): 253–318.

6. Pearson, "Mathematical Contributions to the Theory of Evolution. On the Law of Ancestral Heredity," p. 386.

ard deviation σ, stayed the same each generation. He developed the law *only* as applied to deviations from the mean (of each respective generation) and not to "total values" of characters as Galton had done.

Pearson stated Galton's law with the above modifications in the form

$$k_0 = \frac{1}{2}\frac{\sigma_0}{\sigma_1}k_1 + \frac{1}{4}\frac{\sigma_0}{\sigma_2}k_2 + \frac{1}{8}\frac{\sigma_0}{\sigma_3}k_3 + \frac{1}{16}\frac{\sigma_0}{\sigma_4}k_4 + \ldots ,$$

where k_s was the deviation of the sth mid-parent from the mean of the sth ancestral generation, k_0 the probable deviation from the mean of the offspring of any individual, σ_s the standard deviation of the sth mid-parental generation, and σ_0 the standard deviation of the generation of the offspring.[7]

Having stated Galton's law of ancestral heredity in this form, Pearson went on to demonstrate that one could calculate by complex determinental transformations the correlation between any individual and a parent of his nth ancestral generation from a knowledge of the partial regression between the individual and his mid-nth parent. Since the partial regression coefficients were given in Galton's law by the series $\frac{1}{2}$, $\frac{1}{4}$, $\frac{1}{8}$, . . . one could easily derive the correlation between an individual and any given ancestor. This is why Pearson said one could "predict *a priori* the values of all the correlation coefficients of heredity" by using Galton's law. Pearson concluded that "if Mr. Galton's law can be firmly established, *it is a complete solution, at any rate to a first approximation, of the whole problem of heredity.*"[8]

Pearson did not long remain under the impression that his revision of Galton's law led to "a complete solution . . . of the whole problem of heredity." The problems of fitting alternative inheritance, as in the case of human eye color, into his version of Galton's law soon became apparent. He wrote another paper which he presented to Galton as a second New Year's greeting on 1 January 1900. Pearson began this paper

7. Ibid., p. 387.
8. Ibid., p. 393.

by stating that his revision of Galton's law of ancestral heredity applied only to blended inheritance, and that "blended inheritance certainly does not cover the whole field of heredity." [9] Pearson of course had not mentioned this limitation of Galton's law in his earlier paper. Even Bateson, who admittedly could not follow Pearson's mathematics, saw that Pearson had drastically revised the scope of the law of ancestral heredity between 1898 and 1900. Bateson later commented that "the Law of Ancestral Heredity after the glorious launch in 1898 has been brought home for a complete refit. The top-hamper is cut down and the vessel altogether more manageable; indeed she looks trimmed for most weathers." [10] In the 1900 paper Pearson went on to devise a "law of reversion" which applied only to alternative inheritance. He showed that the law of reversion and the law of ancestral heredity must be carefully distinguished.

Between 1900 and 1903 Pearson made further revisions in the law of ancestral heredity. The reasons for these revisions centered around Galton's series of multiple partial regression coefficients, $\frac{1}{2}$, $\frac{1}{4}$, $\frac{1}{8}$, ... In 1898 Pearson had replaced this rigid series with another containing a single variable to give more flexibility to the law. This series still summed to one. The series of partial regression coefficients was crucially important because by putting them through the elaborate transformation Pearson had devised, the ancestral correlations could be derived. The problem was that these partial regression coefficients could not be observed directly and the correlation coefficients derived from them were therefore suspect.

Pearson was forced to revise his method. Instead of beginning with the partial regression coefficients, he now began with direct measurements of characters from which the correlation coefficients were calculated. Then he reversed his transformation to derive the partial regression coefficients of

9. Karl Pearson, "Mathematical Contributions to the Theory of Evolution. On the Law of Reversion," *Proceedings of the Royal Society* 66 (1900): 141.

10. William Bateson, *Mendel's Principles of Heredity: A Defence* (Cambridge: Cambridge University Press, 1902), p. 112.

the law of ancestral heredity. The process was made simpler because the transformation guaranteed that if the multiple correlation coefficients formed a geometrical series, the multiple regression coefficients would also, and vice versa.

Galton's method was to measure the total regression of offspring on to the mid-parent, then to derive (by questionable means) the ancestral correlations and the partial regression coefficients for his law of ancestral heredity. Pearson's method was to measure the correlation between offspring and parents, grandparents, and great-grandparents. Then he generated a geometrical series to fit these three correlation coefficients and derived by his transformation the geometrical series for the multiple partial regression coefficients of the law of ancestral heredity.

Having made these revisions in Galton's law of ancestral heredity without changing its name, Pearson was of course beset with misunderstandings on the part of biologists. Starting about 1900 he tried to distinguish his own law of ancestral heredity from Galton's. In 1903 he published an article in *Biometrika* entitled "The Law of Ancestral Heredity" [11] in which he tried to elucidate the differences between his interpretation and Galton's. To calculate the correlations between relatives, Galton's method was to measure the correlation (which was the same as the regression under his assumptions) between parent and offspring, then to derive the correlations between grandparent and offspring, etc., by the series $r, r^2, r^3, \ldots$. Pearson's method was to measure the characters of organisms and compute the correlations between parent and offspring, grandparents and offspring, and great-grandparents and offspring. Then he generated a geometrical series which closely fit the observed correlations. In 1903 Pearson was using the series $ar, ar^2, ar^3, \ldots$ with $a = \frac{1}{2}$ and $r = \frac{2}{3}$. The differences are presented in table 3.

Galton derived his series of partial regression coefficients dubiously (see chap. 1). Pearson derived his series by putting the series of ancestral correlations into his transformation,

11. *Biometrika* 2 (1903): 211–28.

TABLE 3

Correlations between Relatives

	Galton's Series	Pearson's Series
Parent and offspring	1/3	1/2
Grandparent and offspring	1/9	1/3
Great-grandparent and offspring	1/27	2/9

TABLE 4

Partial Regression Coefficients between Relatives

	Galton's Series	Pearson's Series
Offspring and mid-parents	1/2	.6244
Offspring and mid-grandparents	1/4	.1988
Offspring and mid-great-grandparents	1/8	.0630

obtaining the partial regression coefficients. The differences may be seen in table 4.

I have described how Galton's interpretation of his own law of ancestral heredity was confused and how Pearson significantly revised the law while at the same time naming it Galton's law of ancestral heredity. Furthermore, Pearson altered his own interpretation of the law several times between 1898 and 1903, using mathematics which was beyond the grasp of most biologists. It is understandable that so much confusion surrounded the meaning and application of the law of ancestral heredity.

Bibliography

Adams, Mark B. "The Founding of Population Genetics: Contributions of the Chetverikov School, 1924–1934." *Journal of the History of Biology* 1 (1968): 23–40.

Bateson, Beatrice. *William Bateson, F. R. S. Naturalist.* Cambridge: Cambridge University Press, 1928.

Bateson, William. *Mendel's Principles of Heredity: A Defence.* Cambridge: Cambridge University Press, 1902.

———. *Scientific Papers of William Bateson.* Edited by R. C. Punnett. 2 vols. Cambridge: Cambridge University Press, 1928.

———, and Saunders, Miss E. R. *Reports to the Evolution Committee of the Royal Society.* Report 1. London: Harrison and Sons, 1902.

Bateson Papers. On microfilm at the Library of the American Philosophical Society, Philadelphia.

Brooks, W. K. *The Law of Heredity: A Study of the Cause of Variation and the Origin of Living Organisms.* Baltimore: John Murphy, 1883.

Castle, W. E. "Mendel's Law of Heredity," *Science,* n.s. 18 (1903): 396–97.

———. "The Laws of Heredity of Galton and Mendel, and Some Laws Governing Race Improvement by Selection." *Proceedings of the American Academy of Arts and Sciences* 39 (1903): 223–42.

———. "The Mutation Theory of Organic Evolution from the Standpoint of Animal Breeding." *Science,* n.s. 21 (1905): 521–25.

———. *Heredity in Relation to Evolution and Animal Breeding.* New York: D. Appleton, 1911.

———. *Genetics and Eugenics.* Cambridge: Harvard University Press, 1916.

———. "Piebald Rats and Selection, a Correction." *The American Naturalist* 53 (1919): 370–76.

———, and MacCurdy, Hansford. *Selection and Cross-Breeding in Relation to the Inheritance of Coat-Pigments and Coat-Patterns in Rats and Guinea-Pigs.* Carnegie Institution of Washington Publication, no. 70. Washington, D.C., 1907.

———, and Phillips, John C. *Piebald Rats and Selection.* Carnegie Institution of Washington Publication, no. 195. Washington, D.C., 1914.

———, and Wright, Sewall. *Studies of Inheritance in Guinea-Pigs and Rats.* Carnegie Institution of Washington Publication, no. 241. Washington, D.C., 1916.

Clark, Ronald W. *JBS: The Life and Work of J. B. S. Haldane.* New York: Coward-McCann, 1969.

Coleman, William. "Bateson and Chromosomes: Conservative Thought in Science." To appear in *Centaurus.*

Darbishire, A. D. "Note on the Results of Crossing Japanese Waltzing Mice with European Albino Races." *Biometrika* 2 (1902): 101-4.

————. "Second Report on the Result of Crossing Japanese Waltzing Mice with European Albino Races." *Biometrika* 2 (1903): 165-73.

————. "Third Report on Hybrids between Waltzing Mice and Albino Races." *Biometrika* 2 (1903): 282-85.

————. "On the Result of Crossing Japanese Waltzing with Albino Mice." *Biometrika* 3 (1904): 1-28.

————. "On the Bearing of Mendelian Principles of Heredity on Current Theories of the Origin of Species." *Manchester Memoirs* 48, no. 24 (1904).

————. "On the Supposed Antagonism of Mendelian to Biometric Theories of Heredity." *Manchester Memoirs* 49, no. 6 (1905).

Darwin, Charles. *The Variation of Plants and Animals under Domestication.* 2 vols. New York: Orange Judd, 1868.

————. *Autobiography.* Edited by Nora Barlow. London: Collins, 1958.

————. *The Origin of Species: A Variorum Text.* Edited by Morse Peckham. Philadelphia: University of Pennsylvania Press, 1959.

————, and Wallace, Alfred Russel. *Evolution by Natural Selection.* Edited by Francis Darwin. Cambridge: University Press, 1958.

Darwin, Francis. *The Life and Letters of Charles Darwin.* 3 vols. London: John Murray, 1887.

De Vries, Hugo. *Intracellular Pangenesis.* Translated by C. Stuart Gager. Chicago: Open Court, 1910.

————. *The Mutation Theory.* Translated by J. B. Farmer and A. D. Darbishire. 2 vols. Chicago: Open Court, 1910.

Dunn, L. C. *A Short History of Genetics.* New York: McGraw-Hill, 1965.

East, Edward M. "The Role of Selection in Plant Breeding." *Popular Science Monthly* 77 (1910): 190-203.

————. "A Mendelian Interpretation of Variation That Is Apparently Continuous." *American Naturalist* 44 (1910): 65-82.

————. "The Role of Reproduction in Evolution." *American Naturalist* 52 (1918): 273-89.

Ellegård, Alvar. *Darwin and the General Reader: The Reception of Darwin's Theory of Evolution in the British Periodical Press, 1859-1872.* Gothenburg Studies in English, vol. 8. Goteburg: Elanders Bohtrycheri Aktiebolag, 1958.

Fisher, R. A. "The Correlation between Relatives on the Supposition of Mendelian Inheritance." *Transactions of the Royal Society of Edinburgh* 52 (1918): 399–433.

————. "On the Dominance Ratio." *Proceedings of the Royal Society of Edinburgh* 42 (1922): 321–41.

————. "On Some Objections to Mimicry Theory: Statistical and Genetic." *Transactions of the Entymological Society of London* 75 (1927): 269–78.

————. "The Possible Modification of the Response of the Wild Type to Recurrent Mutations." *The American Naturalist* 62 (1928): 115–26.

————. *The Genetical Theory of Natural Selection.* Oxford: Clarendon Press, 1930.

————. "Retrospect of the Criticisms of the Theory of Natural Selection." In *Evolution as a Process,* edited by Julian Huxley, A. C. Hardy, and E. B. Ford. New York: Collier Books, 1963.

————, and Ford, E. B. "Variability of Species." *Nature* 118 (1926): 515–16.

Galton, Francis. *Hereditary Genius.* 1869. Reprint of 1892 edition. New York: Meridian Books, 1962.

————. "Experiments in Pangenesis by Breeding from Rabbits." *Proceedings of the Royal Society* 19 (1871): 393–410.

————. "Typical Laws of Heredity." *Journal of the Royal Institution* 8 (1875–77): 282–301.

————. *Natural Inheritance.* London: Macmillan, 1889.

————. *Finger Prints.* London: Macmillan, 1892.

————. "Discontinuity in Evolution." *Mind,* n.s. 3 (1894): 362–72.

————. "The Average Contribution of Each Several Ancestor to the Total Heritage of the Offspring." *Proceedings of the Royal Society* 61 (1897): 401–13.

Haldane, J. B. S. "The Probable Errors of Calculated Linkage Values, and the Most Accurate Method of Determining Gametic from Certain Zygotic Series." *Journal of Genetics* 8 (1919): 291–97.

————. "The Combination of Linkage Values, and the Calculation of Distances between the Loci of Linked Factors." *Journal of Genetics* 8 (1919): 299–309.

————. "A Mathematical Theory of Natural and Artificial Selection." 9 parts. *Transactions* and *Proceedings of the Cambridge Philosophical Society,* 1924–32.

————. "Natural Selection." *Nature* 124 (1929): 444.

————. "The Species Problem in the Light of Genetics." *Nature* 124 (1929): 514–16.

————. "A Note on Fisher's Theory of the Origin of Dominance, and on a Correlation between Dominance and Linkage." *American Naturalist* 64 (1930): 87–90.

————. *The Causes of Evolution.* London: Longmans, Green, 1932.

————, Sprunt, A. D., and Haldane, N. M. "Reduplication in Mice." *Journal of Genetics* 5 (1915): 133–35.

Hanel, Elise. "Vererbung bei Ungeschlechtlicher Fortpflanzung von Hydra Grisea." *Jenaische Zeitschrift* 43 (1908): 322–72.

Hardy, G. H. "Mendelian Proportions in a Mixed Population." *Science,* n.s. 28 (1908): 49–50.

Harris, J. Arthur. "The Biometric Proof of the Pure Line Theory." *American Naturalist* 45 (1911): 346–63.

Hurst, C. C. "On the Inheritance of Coat Colour in Horses." *Proceedings of the Royal Society,* B, 77 (1906): 388–94.

Huxley, Leonard. *Life and Letters of Thomas Henry Huxley.* 2 vols. New York: D. Appleton, 1900.

Huxley, Thomas H. *Darwiniana.* New York: D. Appleton, 1896.

Jennings, H. S. "Heredity, Variation, and Evolution in Protozoa: 2. Heredity and Variation of Size and Form in Paramecium, with Studies of Growth, Environmental Action, and Selection." *Proceedings of the American Philosophical Society* 47 (1908): 393–546.

———. "Heredity and Variation in the Simplest Organisms." *American Naturalist* 43 (1909): 321–37.

———. "Experimental Evidence on the Effectiveness of Selection." *American Naturalist* 44 (1910): 136–45.

———. "Production of Pure Homozygotic Organisms from Heterozygotes by Self-Fertilization." *American Naturalist* 46 (1912): 487–91.

———. "Formulae for the Results of Inbreeding." *American Naturalist* 48 (1914): 693–96.

———. "The Numerical Results of Diverse Systems of Breeding." *Genetics* 1 (1916): 53–89.

———. "Modifying Factors and Multiple Allelomorphs in Relation to the Results of Selection." *American Naturalist* 51 (1917): 301–6.

Johannsen, Wilhelm. *Ueber Erblichkeit in Populationen und in Reinen Linien.* Jena: Gustav Fischer, 1903. Partially translated by Harold Gall and Elga Putschar. In *Selected Readings in Biology for Natural Sciences 3,* pp. 172–215. Chicago: University of Chicago Press, 1955.

———. "Does Hybridization Increase Fluctuating Variability?" In *Report of the Third International Conference on Genetics,* pp. 98–113. London: Spottiswoode, 1907.

Kellogg, Vernon L. *Darwinism To-Day.* New York: Henry Holt, 1907.

Mayr, Ernst. "Where Are We?" *Cold Spring Harbor Symposia on Quantitative Biology* 24 (1959): 1–14.

Mendel, Gregor. *Experiments in Plant Hybridization.* Cambridge: Harvard University Press, 1958.

Morgan, T. H. "For Darwin." *Popular Science Monthly* 74 (1909): 367–80.

———. *A Critique of the Theory of Evolution.* Princeton: Princeton University Press, 1916.

Muller, H. J. "Genetic Variability, Twin Hybrids and Constant Hybrids, in a Case of Balanced Lethal Factors" *Genetics* 2 (1918): 422–99.

Nilsson-Ehle, H. "Kreuzungsuntersuchungen an Hafer und Weizen." *Lunds Universitets Årsskrift*, n.s., ser. 2, vol. 5, no. 2 (1909).

Pearl, Raymond. *Modes of Research in Genetics*. New York: Macmillan, 1915.

————. "The Selection Problem." *American Naturalist* 51 (1917): 65–91.

————, and Surface, Frank. "Is There a Cumulative Effect of Selection?" *Zeitschrift für Induktive Abstammungs und Vererbungslehre* 2 (1909): 257–75.

Pearson, Egon Sharpe. *Karl Pearson*. Cambridge: Cambridge University Press, 1938.

————. "Studies in the History of Probability and Statistics. 20. Some Early Correspondence between W. S. Gosset, R. A. Fisher, and Karl Pearson, with Notes and Comments." *Biometrika* 55 (1968): 445–57.

Pearson, Karl. "Contributions to the Mathematical Theory of Evolution." *Philosophical Transactions of the Royal Society*, A, 185 (1894): 70–110.

————. "Mathematical Contributions to the Theory of Evolution. On the Law of Ancestral Heredity." *Proceedings of the Royal Society* 62 (1898): 386–412.

————. "On the Principle of Homotyposis and Its Relation to Heredity, to the Variability of the Individual, and to That of the Race. Part 1. Homotyposis in the Vegetable Kingdom." *Philosophical Transactions of the Royal Society*, A, 197 (1901): 285–379.

————. "On a Generalized Theory of Alternative Inheritance, with Special Reference to Mendel's Laws." *Philosophical Transactions of the Royal Society*, A, 203 (1904): 53–86.

————. "A Mendelian's View of the Law of Ancestral Inheritance." *Biometrika* 3 (1904): 109–12.

————. "Walter Frank Raphael Weldon, 1860–1906." *Biometrika* 5 (1906): 1–52.

————. "Darwinism, Biometry, and Some Recent Biology." *Biometrika* 7 (1910): 368–85.

————. *The Life, Letters, and Labours of Francis Galton*. 3 vols. Cambridge: Cambridge University Press, 1914–30.

————, and Lee, Alice. "On the Laws of Inheritance in Man. 1. Inheritance of Physical Characters." *Biometrika* 2 (1903): 357–462.

————, and Weldon, W. F. R. "Inheritance in *Phaseolus vulgaris*." *Biometrika* 2 (1903): 499–503.

Punnett, R. C. *Mimicry in Butterflies*. Cambridge: Cambridge University Press, 1915.

————. "Eliminating Feeblemindedness." *Journal of Heredity* 8 (1917): 464–65.

————. "Early Days of Genetics." *Heredity* 4 (1950): 1–10.

Shull, A. Franklin. *Evolution*. New York: McGraw-Hill, 1936.

Stern, Curt. "Wilhelm Weinberg." *Genetics* 44 (1962): 1–5.

Sturtevant, A. H. *An Analysis of the Effects of Selection.* Carnegie Institution of Washington Publication, no. 264. Washington, D.C., 1918.

———. "Thomas Hunt Morgan, 1866–1945." *Biographical Memoirs of the National Academy of Sciences* 33 (1959): 283–325.

———. *A History of Genetics.* New York: Harper and Row, 1965.

Swinburne, R. G. "Galton's Law—Formulation and Development." *Annals of Science* 21 (1965): 15–31.

Vorzimmer, Peter. "Charles Darwin and Blending Inheritance." *Isis* 54 (1963): 371–90.

Warren, Howard C. "Numerical Effects of Natural Selection Acting upon Mendelian Characters." *Genetics* 2 (1917): 305–12.

Weinberg, Wilhelm. "Ueber den Nachweis der Vererbung beim Menschen." *Jahreshefte des Vereins für Vaterländische Naturkunde in Württemburg* 64 (1908): 368–82. English translation in *Papers on Human Genetics,* by Samuel H. Boyer, pp. 4–15. Englewood Cliffs, N.J.: Prentice-Hall, 1963.

———. "Ueber Vererbungsgesetze beim Menschen." *Zeitschrift für Induktive Abstammungs- und Vererbungslehre* 1 (1909): 377–92, 440–60; 2 (1909): 276–330.

———. "Weitere Beiträge zur Theorie der Vererbung." *Archiv für Rassen-und Gesellschafts-Biologie* 7 (1910): 35–49, 169–73.

Weldon, W. F. R. "The Variations Occurring in Certain Decapod Crustacea. 1. *Crangon vulgaris.*" *Proceedings of the Royal Society* 47 (1890): 445–53.

———. "The Study of Animal Variation." *Nature* 50 (1894): 25–26.

———. "Attempt to Measure the Death-rate Due to the Selective Destruction of *Carcinus moenas* with Respect to a Particular Dimension." *Proceedings of the Royal Society* 58 (1895): 360–79.

———. "Remarks on Variation in Animals and Plants." *Proceedings of the Royal Society* 57 (1895): 379–82.

———. "Mendel's Laws of Alternative Inheritance in Peas." *Biometrika* 1 (1902): 228–54.

———. "Professor de Vries on the Origin of Species." *Biometrika* 1 (1902): 365–74.

———. "On the Ambiguity of Mendel's Categories." *Biometrika* 2 (1902): 44–55.

———. "Mr. Bateson's Revisions of Mendel's Theory of Heredity." *Biometrika* 2 (1903): 286–98.

Wentworth, E. N. and Remick, B. L. "Some Breeding Properties of the Generalized Mendelian Population." *Genetics* 1 (1916): 608–16.

Whiting, Phineas T. "Heredity of Bristles in the Common Greenbottle Fly." *American Naturalist* 48 (1914): 343–44.

Wright, Sewall. *An Intensive Study of the Inheritance of Color and of Other Coat Characters in Guinea-Pigs, with Especial*

Reference to Graded Variations. Carnegie Institution of Washington Publication, no. 241. Washington, D.C., 1916.

————. "Color Inheritance in Mammals." 11 parts. *Journal of Heredity* 8 (1917) and 9 (1918).

————. "On the Nature of Size Factors." *Genetics* 3 (1917): 367–74.

————. "The Relative Importance of Heredity and Environment in Determining the Piebald Pattern of Guinea-Pigs." *Proceedings of the National Academy of Sciences* 6 (1920): 320–32.

————. "Correlation and Causation." *Journal of Agricultural Research* 20 (1921): 557–85.

————. "Systems of Mating." *Genetics* 6 (1921): 111–78.

————. "The Effects of Inbreeding and Crossbreeding on Guinea Pigs." *Bulletin of U.S. Department of Agriculture,* nos. 1,090 and 1,121. 1922.

————. "Fisher's Theory of Dominance." *American Naturalist* 63 (1929): 274–79.

————. "The Genetical Theory of Natural Selection. A Review." *Journal of Heredity* 21 (1930): 349–56.

————. "Evolution in Mendelian Populations." *Genetics* 16 (1931): 97–159.

Yule, G. Udny. "Mendel's Laws and Their Probable Relations to Intra-racial Heredity." *New Phytologist* 1 (1902): 193–207, 222–38.

————. "Professor Johannsen's Experiments in Heredity." *New Phytologist* 2 (1903): 235–42.

————. "On the Theory of Inheritance of Quantitative Compound Characters on the Basis of Mendel's Laws—A Preliminary Note." In *Report of the Third International Conference on Genetics,* pp. 140–42. London: Spottiswoode, 1907.

Index

Ancon sheep, 5, 13
Animal Husbandry Division of the United States Department of Agriculture, 155–56, 161

Bailey, Liberty Hyde, 110
Balanoglossus, Bateson's researches on, 36–37
Balfour, Francis, 29–30, 37
Balfour Studentship, 40
Ball, Sir Robert, 140
Bateson, Beatrice, 35–37, 57
Bateson, William, 13, 109; on Balfour's embryological method, 37–38; at British Association, 85–86; and *Chordata,* 37; in Cineraria controversy, 46–48; and Darbishire, 73–80; and Darwinism, 41–43; and de Vries, 66–68; and Evolution Committee, 50–51, 55; and Galton, 42–43; and homotyposis theory, 60–62; and Hurst, 87; and law of ancestral heredity, 54, 185; and Linacre Professorship, 40–41; and materialist theory of heredity, 61; *Materials for the Study of Variation,* 43–45; and mathematics, 36; and Mendelian heredity, 56–57; *Mendel's Principles of Heredity,* 71–72; and Pearson, 63; and Russian trip, 39; sensitivity to criticism, 36; and Weldon, 36–37, 42, 49, 71–72
Bateson, William Henry, 35
Biometrika, foundation of, 62
Bradshaw, Henry, 27–28
Breeding and the Mendelian Discovery (Darbishire), 80
Bridges, C. B., 125
British Association, 1904 meeting of, 85–87
Brooks, W. K., 37–39

Bussey Institution of Harvard, 119, 154–55

Castle, William: and Bateson, 109; and de Vries, 112–13, 121–22; and discontinuous evolution, 110–12; and Hardy-Weinberg law, 132–33; *Heredity in Relation to Evolution and Animal Breeding,* 111–12; and Mendelian factors, 113, 128–29; and Pearson, 110; selection, 111–14, 127–28; and size factors, 156
Catastrophism, in geology, 1–2
Causes of Evolution (Haldane), 169, 174–75
Chambers, Robert, *Vestiges of the Natural History of Creation,* 10
Cineraria, controversy over, 45–48
Clifford, William Kingdom, 26–27
Committee for Conducting Statistical Inquiries into the Measurable Characteristics of Plants and Animals, 32–33, 35, 48–49. *See also* Evolution Committee
Conklin, E. G., 110
Connecticut Agricultural Experiment Station, 119
Continuous evolution: Castle's conversion to, 112–13; Darwin's theory of, 2–3, 5–9; Fisher's theory of, 148–49; Pearson's and Weldon's belief in, 33. *See also* Discontinuous evolution
Continuous variation: in Darwin's theory of natural selection, 5–10; and multiple factor hypothesis, 82–83, 99–100,

Index